Thomas Henry Huxley

An Introduction to the Classification of Animals

Thomas Henry Huxley

An Introduction to the Classification of Animals

ISBN/EAN: 9783337177676

Printed in Europe, USA, Canada, Australia, Japan

Cover: Foto ©ninafisch / pixelio.de

More available books at **www.hansebooks.com**

AN INTRODUCTION

TO THE

CLASSIFICATION OF ANIMALS

BY

THOMAS HENRY HUXLEY, LL.D., F.R.S.,

PROFESSOR OF NATURAL HISTORY IN THE ROYAL SCHOOL OF MINES, AND PROFESSOR OF COMPARATIVE
ANATOMY AND PHYSIOLOGY TO THE ROYAL COLLEGE OF SURGEONS OF ENGLAND.

LONDON:
JOHN CHURCHILL & SONS, NEW BURLINGTON STREET
MDCCCLXIX.

LONDON:
PRINTED BY W CLOWES AND SONS, STAMFORD STREET
AND CHARING CROSS.

PREFACE.

THE present work contains the substance of the six Lectures on the Classification of Animals, which form the first part of my "Lectures on the Elements of Comparative Anatomy," published in 1864.

That book has long been out of print, and I have not had leisure to prepare a new edition of it; but I republish the portion which relates to Classification, because I am told, on good authority, that it is likely to be useful as a text book to lecturers, and students attending lectures, on Comparative Anatomy and Zoology.

I have removed from the original text whatever discussions seemed out of place in a text book, and I have added definitions of all the most important orders of the Animal Kingdom.

My best thanks are due to my friend Dr. PYE-SMITH for the trouble which he has taken in seeing the work through the press, and in preparing the Glossary.

LONDON, MARCH, 1869.

CONTENTS.

CHAPTER I.
ON CLASSIFICATION IN GENERAL 1

CHAPTER II.
THE CHARACTERS OF THE CLASSES OF THE INVERTEBRATA . . 6

CHAPTER III.
THE CHARACTERS OF THE CLASSES OF THE VERTEBRATA . 59

CHAPTER IV.
ON THE ARRANGEMENT OF THE CLASSES INTO LARGER GROUPS 75

CHAPTER V.
THE SUBCLASSES AND ORDERS INTO WHICH THE CLASSES OF THE VERTEBRATA ARE DIVISIBLE 87

CHAPTER VI.
THE ORDERS INTO WHICH THE CLASSES OF THE INVERTEBRATA ARE DIVISIBLE 115

GLOSSARY 131

AN INTRODUCTION

TO THE

CLASSIFICATION OF ANIMALS.

CHAPTER I.

ON CLASSIFICATION IN GENERAL.

BY the classification of any series of objects, is meant the actual, or ideal, arrangement together of those which are like and the separation of those which are unlike; the purpose of this arrangement being to facilitate the operations of the mind in clearly conceiving and retaining in the memory, the characters of the objects in question.

Thus, there may be as many classifications of any series of natural, or of other, bodies, as they have properties or relations to one another, or to other things; or, again, as there are modes in which they may be regarded by the mind: so that, with respect to such classification as we are here concerned with, it might be more proper to speak of *a* classification than of *the* classification of the animal kingdom.

The preparations in the galleries of the Museum of the Royal College of Surgeons are arranged upon the basis laid down by John Hunter, whose original collection was intended to illustrate the modifications which the great physiological apparatuses undergo in the animal series: the classification which he adopted is a classification by organs, and, as such, it is admirably adapted to the needs of the comparative physiologist.

But the student of the geographical distribution of animals, regarding animated creatures, not as diverse modifications of one

B

great physiological mechanism, but in relation to one another, to plants and to telluric conditions, would, with equal propriety, dispose of the contents of a Zoological Museum in a totally different manner; basing his classification, not upon organs, but on distributional assemblages. And the pure palæontologist, looking at life from yet another distinct point of view, would associate animal remains together on neither of these principles, but would group them according to the order of their succession in Time.

Again, that classification which I propose to discuss in the present pages, is different from all of these: it is meant to subserve the comprehension and recollection of the facts of animal structure; and, as such, it is based upon purely structural considerations, and may be designated a Morphological Classification. I shall have to consider animals, not as physiological apparatuses merely; not as related to other forms of life and to climatal conditions; not as successive tenants of the earth; but as fabrics, each of which is built upon a certain plan.

It is possible and conceivable that every animal should have been constructed upon a plan of its own, having no resemblance whatsoever to the plan of any other animal. For any reason we can discover to the contrary, that combination of natural forces which we term Life might have resulted from, or been manifested by, a series of infinitely diverse structures: nor, indeed, would anything in the nature of the case lead us to suspect a community of organization between animals, so different in habit and in appearance, as a porpoise and a gazelle, an eagle and a crocodile, or a butterfly and a lobster. Had animals been thus independently organised, each working out its life by a mechanism peculiar to itself, such a classification as that which is now under contemplation would obviously be impossible; a morphological, or structural, classification plainly implying morphological or structural resemblances in the things classified.

As a matter of fact, however, no such mutual independence of animal forms exists in nature. On the contrary, the different members of the animal kingdom, from the highest to the lowest, are marvellously connected. Every animal has a something in common with all its fellows: much, with many of them;

more, with a few; and, usually, so much with several, that it differs but little from them.

Now, a morphological classification is a statement of these gradations of likeness which are observable in animal structures, and its objects and uses are manifold. In the first place, it strives to throw our knowledge of the facts which underlie, and are the cause of, the similarities discerned into the fewest possible general propositions, subordinated to one another, according to their greater or less degree of generality; and in this way it answers the purpose of a *memoria technica*, without which the mind would be incompetent to grasp and retain the multifarious details of anatomical science.

But there is a second and even more important aspect of morphological classification. Every group in that classification is such in virtue of certain structural characters, which are not only common to the members of the group, but distinguish it from all others; and the statement of these constitutes the definition of the group.

Thus, among animals with vertebræ, the class *Mammalia* is definable as those which have two occipital condyles, with a well-ossified basi-occipital; which have each ramus of the mandible composed of a single piece of bone and articulated with the squamosal element of the skull; and which possess mammæ and non-nucleated red blood-corpuscles.

But this statement of the characters of the class *Mammalia* is something more than an arbitrary definition. It does not merely mean that naturalists agree to call such and such animals *Mammalia*: but it expresses, firstly, a generalization based upon, and constantly verified by, very wide experience; and, secondly, a belief arising out of that generalization. The generalization is that, in nature, the structures mentioned are always found associated together: the belief is, that they always have been, and always will be, found so associated. In other words, the definition of the class *Mammalia* is a statement of a law of correlation, or coexistence, of animal structures, from which the most important conclusions are deducible.

For example: if a fragmentary fossil be discovered, consisting of no more than a ramus of a mandible and that part of the skull with which it articulated, a knowledge of this law may enable the palæontologist to affirm, with great confidence, that the animal of which it formed a part suckled its young and had non-nucleated red blood-corpuscles; and to predict that, should the back part of that skull be discovered, it will exhibit two occipital condyles and a well-ossified basi-occipital bone.

Deductions of this kind, such as that made by Cuvier in the famous case of the fossil opossum of Montmartre, have often been verified, and are well calculated to impress the vulgar imagination; so that they have taken rank as the triumphs of the anatomist. But it should carefully be borne in mind, that, like all merely empirical laws, which rest upon a comparatively narrow observational basis, the reasoning from them may at any time break down. If Cuvier, for example, had had to do with a fossil *Thylacinus* instead of a fossil Opossum, he would not have found the marsupial bones, though the inflected angle of the jaw would have been obvious enough. And so, though, practically, any one who met with a characteristically mammalian jaw would be justified in expecting to find the characteristically mammalian occiput associated with it; yet, he would be a bold man indeed, who should strictly assert the belief which is implied in this expectation, viz., that at no period of the world's history did animals exist which combined a mammalian occiput with a reptilian jaw, or *vice versâ*.

Not that it is to be supposed that the correlations of structure expressed by these empirical laws are in any sense accidental, or other than links in the general chain of causes and effects. Doubtless there is some very good reason why the characteristic occiput of a Mammal should be found in association with mammæ and non-nucleated blood-corpuscles; but it is one thing to admit the causal connection of these phenomena with one another, or with some third; and another thing to affirm that we have any knowledge of that causal connexion, or that physiological science, in its present state, furnishes us with any means of reasoning from the one to the other.

Cuvier, the more servile of whose imitators are fond of citing his mistaken doctrines as to the nature of the methods of palæontology against the conclusions of logic and of common sense, has put this so strongly that I cannot refrain from quoting his words.*

"But I doubt if any one would have divined, if untaught by observation, that all ruminants have the foot cleft, and that they alone have it. I doubt if any one would have divined that there are frontal horns only in this class: that those among them which have sharp canines for the most part lack horns.

"However, since these relations are constant, they must have some sufficient cause; but since we are ignorant of it, we must make good the defect of the theory by means of observation: it enables us to establish empirical laws, which become almost as certain as rational laws, when they rest on sufficiently repeated observations; so that now, whoso sees merely the print of a cleft foot may conclude that the animal which left this impression ruminated, and this conclusion is as certain as any other in physics or morals. This footprint alone, then, yields to him who observes it, the form of the teeth, the form of the jaws, the form of the vertebræ, the form of all the bones of the legs, of the thighs, of the shoulders, and of the pelvis of the animal which has passed by: it is a surer mark than all those of Zadig."

* 'Ossemens fossiles,' ed. 4me, tomo 1^r, p. 184.

CHAPTER II.

THE CHARACTERS OF THE CLASSES OF THE INVERTEBRATA.

MORPHOLOGICAL classification, then, acquires its highest importance as a statement of the empirical laws of the correlation of structures; and its value is in proportion to the precision and the comprehensiveness with which those laws, the definitions of the groups adopted in the classification, are stated. So that, in attempting to arrive at clear notions concerning classification, the first point is to ascertain whether any, and if so, what groups of animals can be established, the members of which shall be at once united together and separated from those of all other groups, by well-defined structural characters. And it will be most convenient to commence the inquiry with groups of that value which are commonly called CLASSES, and which are enumerated in an order and arrangement, the purpose of which will appear more fully by and by, in the following table.

TABLE OF THE CLASSES OF THE ANIMAL KINGDOM.

The Limits of the Four Cuvierian Sub-Kingdoms are indicated by the Brackets and Dotted Line.

RADIATA.

Gregarinida.	*Infusoria.*	*Scolecida* (?).	
Rhizopoda.		*Echinodermata.*	
Radiolaria.			
Spongida.		*Chœtognatha.*	
		Annelida.	
Hydrozoa.		*Crustacea.*	ARTICULATA.
Actinozoa.		*Arachnida.*	
		Myriapoda.	
Polyzoa.		*Insecta.*	

Brachiopoda.
Ascidioida.

Lamellibranchiata.

Branchiogasteropoda.
Pulmogasteropoda.
Pteropoda.
Cephalopoda.

} Mollusca.

Pisces.
Amphibia.
Reptilia.
Aves.
Mammalia.

} Vertebrata.

It is not necessary for my purpose that the groups which are named on the preceding table should be absolutely and precisely equivalent one to another; it is sufficient that the sum of them is the whole of the Animal Kingdom, and that each of them embraces one of the principal types, or plans of modification, of animal form; so that, if we have a precise knowledge of that which constitutes the typical structure of each of these groups, we shall have, so far, an exhaustive knowledge of the Animal Kingdom.

I shall endeavour, then, to define—or, where definition is not yet possible, to describe a typical example of—these various groups. Subsequently, I shall take up some of those further classificatory questions which are open to discussion; inquiring how far we can group these classes into larger assemblages, with definite and constant characters; and, on the other hand, how far the classes can be broken up into well-defined sub-classes and orders. But the essential matter, in the first place, is to be quite clear about the different classes, and to have a distinct knowledge of all the sharply-definable modifications of animal structure which are discernible in the animal kingdom.

I. The Gregarinida.

These are among the simplest animal forms of which we have any knowledge. They are the inhabitants of the bodies, for the most part, of invertebrate, but also of vertebrate, animals; and they are commonly to be found in abundance in the alimentary canal of the common cockroach, and in earth-

8 INTRODUCTION TO CLASSIFICATION.

worms. They are all microscopic, and any one of them, leaving minor modifications aside, may be said to consist of a sac, composed of a more or less structureless, not very well-defined membrane, containing a soft semi-fluid substance, in the midst, or at one end, of which lies a delicate vesicle; in the centre of the latter is a more solid particle. (Fig. 1, A.) No doubt many persons will be struck with the close resemblance of the structure of this body to that which is possessed by an ovum.

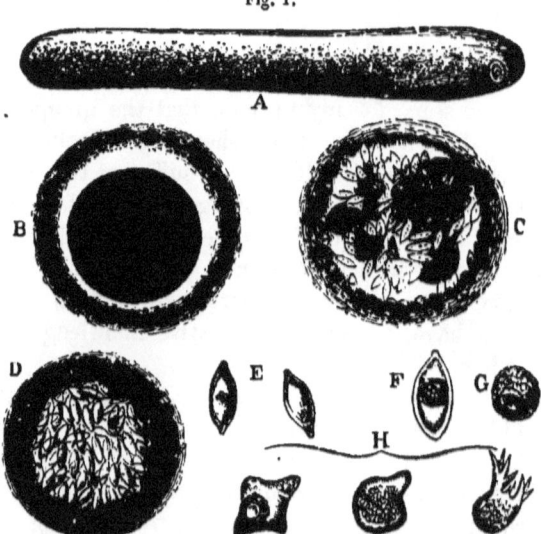

Fig. 1.—A, *Gregarina* of the earthworm (after Lieberkühn); B, encysted; C, D, with the contents divided into psuedo-navicellæ; E, F, free pseudo-navicellæ; G, H, free amœbiform contents of the latter.

You might take the more solid particle to be the representative of the germinal spot, and the vesicle to be that of the germinal vesicle; while the semi-fluid sarcodic contents might be regarded as the yelk, and the outer membrane as the vitelline membrane. I do not wish to strain the analogy too far, but it is, at any rate, interesting to observe this close morphological resemblance between one of the lowest of animals and that form in which all the higher animals commence their existence. It is a very remarkable characteristic of this group, that there is no separation of the body into distinct layers, or into cellular elements. The *Gregarinida* are devoid of mouths and of digestive apparatus, living entirely by imbibition of the juices of the animal in whose intestine, or body cavity, they are contained. The most conspicuous of those phenomena, which we ordinarily regard as signs of life, which they exhibit, is a certain contraction and expansion along different diameters, the body slowly narrowing, and then lengthening, in various directions. Under certain circumstances (though the conditions of the change are not thoroughly understood), it is observed that one of these *Gregarinida*, whatever its form may be, will convert itself into a well-rounded sac, the outer membrane ceasing to exhibit any longer those movements of which I spoke, and becoming coated by a structureless investment, or "cyst" (Fig. 1, B).

The substance of the body contained within the cyst next undergoes a singular change. The central nucleus and the vesicle disappear; after a time, the mass breaks up into a series of rounded portions and, then, each of those rounded portions elongates, and, becoming slightly pointed at each end, constitutes a little body which has been called a "*Pseudo-navicella,*" from its resemblance to the Diatomaceous *Navicula* or *Navicella* (Fig. 1, C, D). Next, the capsule bursts and the *Pseudo-navicellæ* (Fig. 1, E, F) are scattered and passed out of the body of the animal which they inhabit. Though, of course, a great number of them are destroyed, some, at any rate, are devoured by other animals; and, when that is the case, the little particle of protein substance which is enclosed within the *Pseudo-navicella* is set free from its shell, and exhibits much more lively movements than before, thrusting out processes in various directions, and drawing them in again, and, in fact, closely resembling one of those animalcules which have been called *Amœbæ* (Fig. 1, H). The young Amœbiform *Gregarina* grows, increases in size, and at length assumes the structure which it had at first. That, in substance, is all that we know of this lowest division of animal life. But it will be observed, there is a hiatus in our knowledge. We cannot say that we know the whole nature and mode of existence of this, or any other animal, until we have traced it to its sexual state; but, at present, we know nothing whatever of this condition among the *Gregarinæ;* so that in reasoning about them we must always exercise a certain reticence, not knowing how far we may have to modify our opinions by the discovery of the sexual state hereafter.

The process of becoming encysted, preceded or accompanied very often by the mutual apposition of two *Gregarinæ*, was formerly imagined to correspond with what is termed among plants "conjugation,"—a process which in some cases, at any rate, appears to be of a sexual nature. But the discovery that a single *Gregarina* may become encysted and break up into *Pseudo-navicellæ*, seems to negative this analogy.

II. THE RHIZOPODA.

It seems difficult to imagine a stage of organization lower than that of *Gregarinida*, and yet many of the *Rhizopoda* are still simpler (Fig. 2). Nor is there any group of the animal kingdom which more admirably illustrates a very well-founded doctrine, and one which was often advocated by John Hunter, that life is the cause and not the consequence of organization; for, in these lowest forms of animal life, there is absolutely nothing worthy of the name of organization to be discovered by the microscopist, though assisted by the beautiful instruments that are now constructed. In the substance of many of these creatures, nothing is to be discerned but a mass of jelly, which might be represented by a little particle of thin glue. Not that it corresponds with the latter in composition, but it has that texture and sort of aspect; it is structureless and organless, and without definitely-formed parts. Nevertheless, it possesses all the essential properties and characters of vitality; it is produced from a body like itself; it is capable of assimilating nourishment, and of exerting movements. Nay, more, it can produce a shell; a structure, in many cases, of extraordinary complexity and most singular beauty (Fig. 2, D).

That this particle of jelly is capable of guiding physical forces in such a manner as to give rise to those exquisite and almost mathematically-arranged structures—being itself structureless and without permanent distinction or separation of parts —is, to my mind, a fact of the profoundest significance.

Though a Rhizopod is not permanently organized, however, it can hardly be said to be devoid of organs; for the name of the group is derived from the power which these animals possess of throwing out processes of their substance, which are called "pseudopodia," and are sometimes very slender and of great length (Fig. 2, E), sometimes broad and lobe-like (Fig. 2, A).

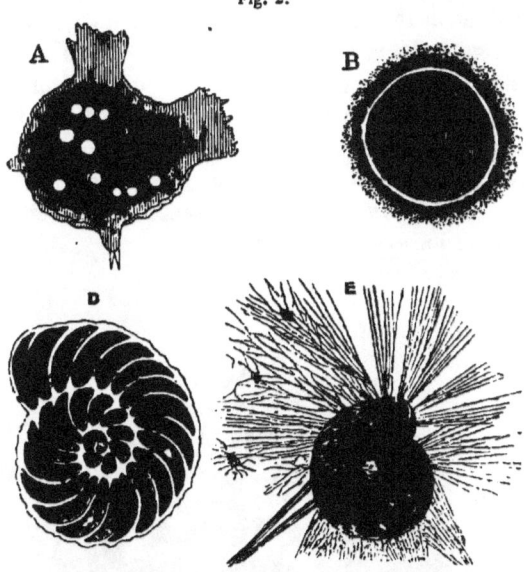

Fig. 2.

Fig. 2.—A, B, Free and encysted conditions of an *Amœba* (after Auerbach); E, a Foraminifer (*Rotalia*) with extended pseudopodia; D, its shell in section (after Schulze).

These processes may flow into one another, so as to form a network, and they may, commonly, be thrust out from any part of the body and retracted into it again.

If you watch one of these animals alive, you see it thrusting out, first one and then another of its pseudopodia, exhibiting changes of form comparable to those which the colourless corpuscles of the human blood present. The movements of these Rhizopods are quite of the same character; they are rapid, extensive, and effect locomotion. The creature also feeds itself by means of its pseudopodia, which attach themselves to nutritive particles, and then draw them into the substance of the body. There is neither ingestive nor egestive aperture, neither special motor nor prehensile organs, but the pseudopodia perform each function as it may be required.

But here, again, we labour under an imperfection of knowledge. For, although it is quite certain that the *Rhizopoda* may multiply by division of their substance—in a way somewhat analogous to that which I detailed when speaking of the *Gregarinida*—yet, as in that case, we have no knowledge of any true sexual process. It is a most remarkable circumstance that though these animals are abundant, and are constantly under observation, we are still in doubt upon that essential point,—still uncertain whether there may not be some phase in the cycle of vital phenomena of the *Rhizopoda* with which we are unacquainted; and, under these circumstances, a perfect definition of the class cannot even be attempted.

Fig. 3.

Fig. 3.—*Sphærozoum ovodimare* (after Haeckel), one of the complex *Radiolaria*.

III. THE RADIOLARIA.

The simple forms of this group consist of microscopic masses of sarcode or protoplasm, from which long slender pseudopodia, which may unite into reticulations, protrude. This protoplasmic substance contains a sac in which are inclosed cellæform bodies, fat globules, coloured granules or crystals, with more or less protoplasm. Very generally, numerous yellow corpuscles, which multiply by fission, are scattered through the superficial protoplasm. To these parts a skeleton may be added, consisting of spicula (which may be loose, or united into a shell imbedded in the superficial protoplasm), or of rods, which meet in the middle of the sac. The skeleton is usually silicified. The more complex forms consist of aggregations of the simpler, which may inclose " vacuoles " or spaces full of water, as in *Sphærozoum* (Fig. 3). No sexual process has been observed in any Radiolarian.

The siliceous skeletons of some of the *Radiolaria* are known under the name of *Polycistineæ*, and, like the skeletons of the *Foraminifera*, they enter largely into the formation of some strata of the earth's crust.

IV. THE SPONGIDA.

Multitudinous forms of sponges exist in both salt and fresh waters. Up to the last few years we were in the same case, with respect to this class, as with the *Gregarinida*, the *Rhizopoda*, and the *Radiolaria*. Some zoologists even have been anxious to relegate the sponges to the vegetable kingdom; but the botanists, who understood their business, refused to have anything to do with the intruders. And the botanists were quite right; for the discoveries of late years have not left the slightest doubt that the sponges are animal organisms, and animal organisms, too, of a very considerable amount of complexity, if we may regard as complex a structure which results from the building up and massing together of a number of similar parts.

The great majority of the sponges form a skeleton, which is composed of fibres of a horny texture, strengthened by needles, or spicula, of siliceous, or of calcareous, matter; and this framework is so connected together as to form a kind of fibrous skeleton. This, however, is not the essential part of the animal, which is to be sought in that gelatinous substance, which invests the fibres of the skeleton during life, and is traversed by canals which open upon the surface of the sponge, directly or indirectly, by many minute, and fewer large, apertures.

If I may reduce a sponge to its simplest expression—taking the common *Spongilla*, for example, of our fresh waters,—the structure—removing all complexities, and not troubling ourselves with the skeleton, because that has nothing to do with what we are now considering—may be represented by the diagram (A, Fig. 4). There is a thin superficial layer (*a*) formed entirely of a number of the so-called sponge particles, or ultimate components of the living substance of the sponge, each of which is similar to an *Amœba*, and contains a nucleus. These are all conjoined in a single layer, so as to form a continuous lamellar membrane, which constitutes the outer and superficial layer of the body. Beneath this is a wide cavity, communicating with the exterior by means of minute holes in the superficial layer (*b*), and, of course, filled with water. The cavity separates the superficial layer of the sponge from its deeper substance, which is of the same character as the superficial layer, being made up of a number of aggregated sponge particles, each of which has a nucleus, and is competent to throw out numerous pseudopodial prolongations if detached. While the living sponge is contained in water, a great number of currents of water set in to the wide cavity beneath *a, a*, through the minute apertures (*b*), which have thence been termed " inhalent."

In the floor of the cavity there are a number of apertures which lead into canals ramifying in the deep layer, and eventually ending in the floors of certain comparatively lofty funnels, or craters. The top of each of these presents one of those larger and less numerous apertures, which have been referred to as existing on the surface of the sponge, and which are fitly termed " exhalent" apertures. For it has been discovered that strong, though minute, currents of water are constantly flowing out of these large apertures; being fed by the currents which as constantly set in, by the small apertures and through the superficial cavity, into the canals of the deeper substance. The cause of this very singular system of currents is the existence of vibratile

THE SPONGIDA. 15

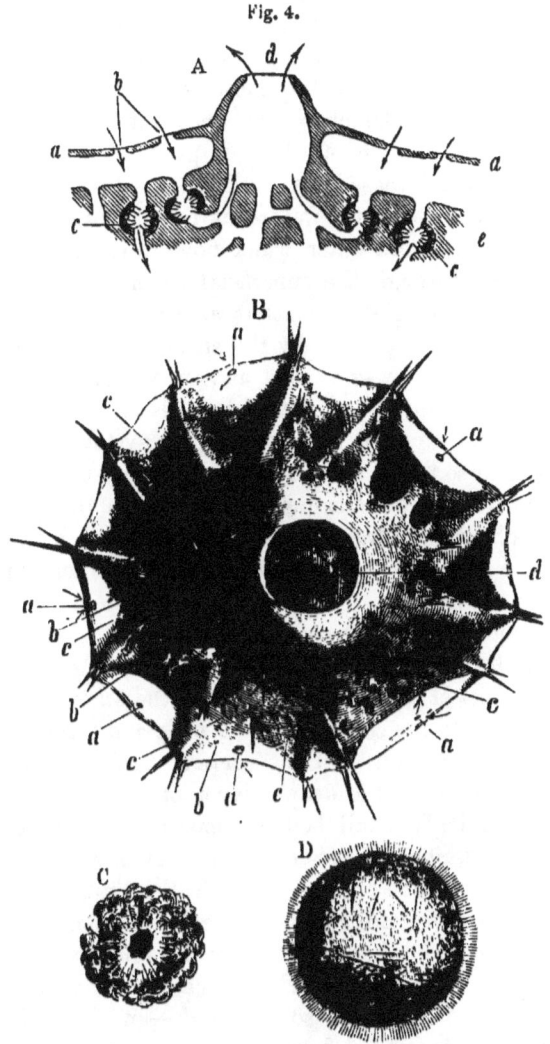

Fig. 4.—A, Hypothetical section of a *Spongilla*; *a*, superficial layer; *b*, inhalent apertures; *c*, ciliated chambers; *d*, an exhalent aperture; *e*, deeper substance of the sponge. The arrows indicate the direction of the currents. B, A small sponge with a single exhalent aperture, seen from above (after Lieberkühn); *a*, inhalent apertures; *c*, ciliated chambers; *d*, exhalent aperture. C, A ciliated chamber. D, A free-swimming ciliated embryo.

cilia in the water-passages of the sponge, but it is only quite recently that the precise nature of the arrangement of the apparatus which gives rise to these currents, has been made out. The canals which enter the deep substance of the sponge become dilated into spheroidal chambers, lined with sponge particles (Fig. 4, A, c, C), each of which is provided with a vibratile cilium; and as all these cilia work in one direction—towards the crater—they sweep the water out in that direction, and its place is taken by fresh water, which flows in through the small apertures and through the superficial chamber. The currents of water carry along such matters as are suspended in them; and these are appropriated by the sponge particles lining the passages, in just the same way as any one of the *Rhizopoda* appropriates the particles of food it finds in the water about itself. So that we must not compare this system of apertures and canals to so many mouths and intestines; but the sponge represents a kind of subaqueous city, where the people are arranged about the streets and roads, in such a manner, that each can easily appropriate his food from the water as it passes along.

Two reproductive processes are known to occur in the sponges: the one of them, asexual, corresponding with the encysting process of the *Gregarinida;* and the other, truly sexual, and answering to the congress of the male and female elements in the higher animals. In the common fresh-water *Spongilla*, towards the autumn, the deeper layer of the sponge becomes full of exceedingly small bodies, sometimes called "seeds" or "gemmules," which are spheroidal, and have, at one point, an opening. Every one of these bags—in the walls of which are arranged a great number of very singular spicula, each resembling two toothed wheels joined by an axle—is, in point of fact, a mass of sponge particles which has set itself apart—gone into winter quarters, so to speak—and becoming quite quiescent, encysts itself and remains still. The whole *Spongilla* dies down, and the seeds, inclosed in their case, remain uninjured through the winter. When the spring arrives, the encysted masses within the "seed," stimulated by the altered temperature of the water, creep out of their nests, and straightway grow up into *Spongillæ* like that from which they proceeded.

But there is, in addition, a true sexual process, which goes on during the summer months. Individual sponge particles become quiescent, and take on the character of ova; while, in other parts, particular sponge particles fill with granules, the latter eventually becoming converted into spermatozoa.

These sacs burst and some of the spermatozoa, coming into contact with the ova, impregnate them. The ova develop and grow into ciliated germs (D, Fig. 4), which make their way out, and, after swimming about for a while, settle themselves down and grow up into *Spongillæ*.

Now that we know the whole cycle of the life of the sponges, and the characters which may be demonstrated to be common to the whole of this important and remarkable class, I do not think any one who is acquainted with the organization or the functions of plants will be inclined to admit that the *Spongida* have the slightest real affinity with any division of the vegetable kingdom.

V. THE INFUSORIA.

Although the *Infusoria* have been favourite studies for many years, it is only quite recently that the anatomy of these animals has been satisfactorily made out.

The different species of the infusorial genus *Paramœcium* are very common among the microscopic inhabitants of our fresh waters, swimming about by means of the vibratile cilia with which the whole surface of their bodies is covered; and the structure which essentially characterises these animals is probably that which is common to the whole of the *Infusoria*, so that an account of the leading structural features of *Paramœcium* is, in effect, a definition of those of the group.

Imagine a delicate, slipper-shaped body inclosed within a structureless membrane, or *cuticula*, which is formed as an excretion upon its outer surface. At one point (Fig. 5, B *a*) the body exhibits a slight depression, leading into a sort of little funnel (*b c*) coated by a continuation of the same cuticular investment,

C

which stops short at the bottom of the funnel. The whole of the bag formed by the cuticula is lined by a soft layer of gelatinous matter, or "sarcode," which is called the "cortical" layer (Fig. 5, A *a*); while inside that, and passing into it quite gradually, there being no sharp line of demarcation between the two, is a semi-fluid substance, which occupies the whole of the central region of the body. Neither in the cuticle, the cortical layer, nor the central substance, has any anatomist yet discovered a differentiation into cellular layers, nor any trace of

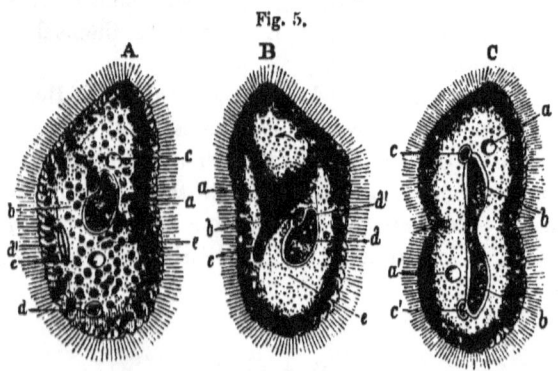

Fig. 5.—*Paramœcium bursaria* (after Stein): A, The animal viewed from the dorsal side; *a*, cortical layer of the body; *b*, "nucleus;" *c*, contractile chamber; *d d'*, matters taken in as food; *e*, chlorophyll granules.

B, The animal viewed from the ventral side; *a*, depression leading to *b*, mouth; *c*, gullet; *d*, "nucleus;" *d'*, "nucleolus;" *e*, central sarcode. In both these figures the arrows indicate the direction of the circulation of the sarcode.

C, *Paramœcium* dividing transversely; *a a'*, contractile spaces; *b b'*, "nucleus" dividing; *c c'*, "nucleoli."

that histological composition which we meet with in the tissues of the higher animals; so that here is another case of complex vital phenomena proceeding from a substance which, in a histological sense, is structureless.

At two points of the body (Fig. 5, A, *c, c*) the substance of the cortical layer exhibits a remarkable power of contraction and dilatation. If you watch one of those points, the sarcode suddenly seems to open like a window, and, for a while, a clear space is visible, which then, quite suddenly, shuts again. After a little time the same diastole and systole are repeated. As the systole takes place, it is possible, occasionally, to discern certain radiating canals, which extend from the cavities into the surrounding sarcode, and disappear again before diastole occurs. There is no doubt that the clear space is a chamber filled with fluid in the cortical layer; and since good observers maintain that there is an aperture of communication, through the cuticula, between the 'contractile chamber' and the exterior, this fluid can be little more than water. Perhaps the whole should be regarded as a respiratory or secretory mechanism: in one shape or another, it is eminently characteristic of the *Infusoria*. Besides this singular apparatus, there lies embedded in another part of the cortical layer a solid mass, of an elongated oval shape (Fig. 5, A, B, d), which has been called the "nucleus," though it must be carefully distinguished from the "nucleus" of a cell. Upon one side of this, and, as it were, stuck on to it, is a little rounded body (Fig. 5, B, d'), which has received the name of the "nucleolus." The animal swims about, driven by the vibration of its cilia, and whatever nutriment may be floating in the water is appropriated by means of the current which is caused to set continually into the short gullet by the cilia which line that tube.

But it is a singular circumstance, that these animals have an alimentary canal consisting of a mere gullet, open at the bottom, and leading into no stomach or intestine, but opening directly into the soft central mass of sarcode. The nutritious matters passing down the gullet, and then into the central more fluid substance, become surrounded by spheroids of clear liquid (Fig. 5, A, d), consisting apparently of the water swallowed with them, so that a well-fed *Paramœcium* exhibits a number of cavities, each containing a little mass of nutritious particles. Hence formerly arose the notion that these animals possess a number of stomachs. It was not unnaturally imagined that each of the cavities in question was a distinct stomach; but it has since been discovered that the outer layer of the sarcode is, by means of some unknown mechanism, kept in a state of constant rotation; so that the supposed stomachs may be seen to undergo a regular circulation up one side of the body and down the other. And this circumstance, if there were no other arguments on the same side, is sufficient to negative the supposition that the food-containing spaces are stomachs; for it is impossible to imagine any kind of anatomical arrangement which shall permit true dilatations of an alimentary canal to rotate in any such manner. Fæcal matters are extruded from an anus, which is situated not far from the mouth, but is invisible when not in use. It is an interesting and important character of the *Infusoria*, in general, that, under some circumstances, they become quiescent and throw out a structureless cyst around their bodies. The *Infusorium* then not unfrequently divides and subdivides, and, the cyst bursting, gives rise to a number of separate *Infusoria*.

The remarkable powers of multiplication by fission which many of the group exhibit are well known; but within the last few years the investigations of Müller, Balbiani, Stein, and others, have led them to believe that these minute creatures possess a true process of sexual multiplication, and that the sexual organs are those which have been denominated "nucleus" and "nucleolus." The nucleus is considered to be the true ovary—the nucleolus, the testis, in *Paramœcium*. But further information is required before this interpretation can be finally accepted. All that can be said to be made out with perfect certainty is, that occasionally vibrio-like rods (the supposed spermatozoa) are seen in the enlarged and modified nucleolus and nucleus; and that the nucleus may under certain circumstances give rise to germs by fission.

A process of conjugation has been observed in many *Infusoria*.

In giving an account of the preceding groups, I have substituted for a definition of each class a description of the structure of some particular member of that class, or of the organic features which are most obviously characteristic of the class; because, in hardly any of those groups has the structure of many, and widely different, members been thoroughly and exhaustively worked out.

I entertain little doubt, however, that the main features of the description of *Spongilla* might substantially be taken as a definition of the *Spongida*, and those of the description of *Paramœcium*, as a definition of the *Infusoria*. On the other hand, we possess no such complete knowledge of the vital cycle of any *Gregarina*, *Rhizopod*, or *Radiolarian*; and neither description nor definition of the corresponding classes, of a thoroughly satisfactory kind, is attainable.

No such difficulties beset us in studying the next class, which embraces the Hydroid polypes and the *Medusæ*, and which may be defined with as much precision as any group in the Animal Kingdom.

VI. The Hydrozoa.

All the *Hydrozoa* exhibit a definite histological structure, their tissues primarily presenting that kind of organization which has been called cellular. Again, the body always exhibits a separation into at least two distinct layers of tissue—an outer and an inner—which have been termed, respectively, *ectoderm* and *endoderm*. The endoderm is that layer which lines the inner cavities of the body, from the mouth inwards; the ectoderm is that which forms its external covering.

These two layers are shown in the accompanying diagrammatic sections of the leading forms of *Hydrozoa*, the ectoderm being represented by the thin line with the adjacent clear space, the endoderm by the thick dark line (Fig. 6).

A third distinctive character of the *Hydrozoa* is, that the digestive cavity communicates directly, by a wide aperture, with the general cavity of the body; the one, in fact, passing by direct continuity into the other. Furthermore, the digestive sac is not in any way included in the substance of the rest of the body, but stands out independently, so that the outer wall of the digestive cavity is in direct contact with the water in which the animal lives, and there is no perivisceral chamber. The like is true of the reproductive organs, which may vary very much in form, but have the common peculiarity of being developed as outward processes of the body wall, so that their external surfaces are directly in contact with the surrounding medium.

The majority of these animals seize their prey by means of tentacula developed either around the mouth, or from the walls of the digestive cavity, or from the body wall; and these tentacles, as well as other parts of the body, are provided with those peculiar weapons of offence which have been termed "thread-cells."

Fig. 6.

Fig. 6.—Diagrams illustrative of the mutual relations of the *Hydrozoa*.—1. *Hydra*. 2. Sertularian. 3. Diphyes. 4. Physophorid. 5. *Lucernaria*. *a*. Ectocyst. *b*. Endocyst. *c*. Their enclosed cavity.
P. Tentacles. N. Natatorial organ. T. Cœnosarc. B. Bract. C. Cell. S. Polypite or digestive cavity. G. Reproductive organ. A. Air vesicle. F. Float.
I., II., III., IV. represent the successive stages of development of a Medusiform zooid or reproductive organ.

VII. THE ACTINOZOA.

This class contains those animals which are familiar to us as Sea-anemones and Coral-polypes, by the latter of which, in many parts of the world, those huge reefs, which are so well known to navigators, are constructed. It embraces the Sea-pens and the Red coral, and those creatures which are known to us under the names of *Beroë, Cydippe, Pleurobrachia,* &c., transparent, beautifully symmetrical, free-swimming animals, provided with eight rows of longitudinally-disposed large cilia. In all these animals we find a great uniformity of structure, and their plan of construction is quite as readily definable as that of the preceding class, with which they exhibit a close affinity. Like the majority of the *Hydrozoa*, most *Actinozoa* have their mouths surrounded by tentacles; and there is the same primary distinction of the body into two cellular layers—the ectoderm and the endoderm—though, in the adult forms of the more highly organized *Actinozoa*, these primitive layers become further differentiated into bundles of definitely disposed muscular fibres, and even into nerves and ganglia.

As in the *Hydrozoa*, again, the alimentary canal communicates freely, and by a wide aperture, with the general cavity of the body; but the whole of the *Actinozoa*, polype-like as they are in external appearance, differ from the *Hydrozoa* by a very important further progress towards complexity. We found that in the *Hydrozoa* the digestive cavity was completely outside the general cavity of the body, the digestive portion of the organism being continued into, and not in any way contained within, the part which surrounds the general cavity. But if you make a vertical section of a sea-anemone (Fig. 7), you will find that the alimentary cavity—as freely open at the bottom as in the *Hydrozoa*—is enclosed within a part of the body which contains a prolongation of the general cavity. If you could suppose the stomach of a *Hydrozoon* thrust into that part of the body with which it is continuous, so that the walls of the body should rise round it and form a sort of outside case, containing a prolongation of the general cavity, the *Hydrozoon* would be converted into an *Actinozoon*.

The prolongation of the general cavity thus produced, which, as it surrounds the chief viscus, may be termed the " perivisceral cavity " (d), receives the products of digestion mixed with much sea-water; and the nutritive fluid, which fills the perivisceral cavity and its ramifications, plays the same part as the blood of the more highly organized animals. The gastric chamber of the *Actinozoa* does not lie free in the interior of the body, but is connected to the sides of it by means of membranous partitions, the so-called "mesenteries" (*f*), which pass radially from the stomach to the side walls of the body, and so divide the "peri-

Fig. 7.

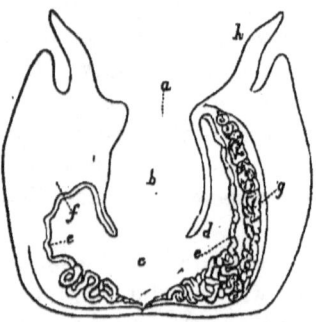

Fig. 7.—Perpendicular section of *Actinia holsatica* (after Frey and Leuckart); *a*, mouth; *b*, alimentary cavity; *c*, common cavity; *d*, intermesenteric chambers; *e*, cord containing thread-cells at the edge; *f*, the mesentery; *g*, reproductive organ; *h*, tentacle.

visceral cavity" into a number of chambers, which communicate with the bases of the tentacles. In the whole of the *Hydrozoa* the reproductive organs were attached to the exterior of the body, and projected from it. In the whole of the *Actinozoa*, on the other hand, the reproductive organs (of which both sexes are frequently combined in the same individual) are internal, inasmuch as they are situated in the substance of the mesenteries (*g*).

These are the universal and distinctive characters of the *Actinozoa*. That some are simple and some are compound organisms; that some are fixed and some free swimmers; that many are soft, while a great number are provided with very dense skeletons; that some possess a rudimentary nervous system, while the majority have as yet afforded no trace of any such structure—are secondary circumstances in no way affecting the problem before us, which is, to find a diagnostic definition of the group.

VIII. THE POLYZOA.

Notwithstanding the invariably minute size of the organisms which constitute this class, they exhibit a very great advance in complexity of structure. In such a compound *Polyzoon* as the Sea-mat, or *Flustra*, the entire surface of the foliaceous expansion, on being examined by the microscope, will be found to be beset with an infinitude of minute apertures leading into little chambers, out of each of which, when the animal was living and active, multitudes of little creatures might be seen protruding the oral extremities of their bodies. The ends of the branches of the freshwater genus *Plumatella*, represented in Fig. 8, present a similar spectacle. Each mouth is surrounded by a circlet of tentacles; and, as every tentacle is fringed with long and active vibratile cilia, lashing the water towards the mouth, hundreds and thousands of little Maelströms are created, each tending to suck down such nutritious bodies, living or dead, as come within its range. The mouth (Fig. 9) leads into a long and wide pharyngeal and œsophageal tube, which opens, below, into a definite stomach. From this is continued a distinct intestine, which

Fig. 8.

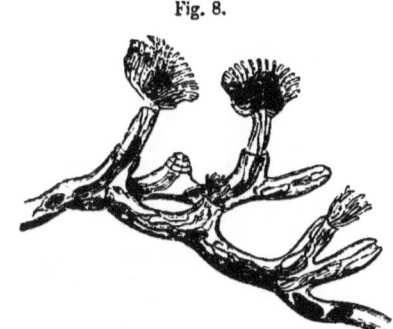

Fig. 8.—*Plumatella repens*, a fresh-water Polyzoon, magnified (after Allman).

bends upon itself towards the oral end of the body, so as to form a sharp angle, and then terminates upon the outer surface near the mouth; so that we have here, for the first time in our ascending survey of the Animal Kingdom, an animal possessing a complete intestine, not only structurally separated from the general substance of the body, and provided with permanent apertures, as in the *Hydrozoa* and *Actinozoa*, but completely shut off from the perivisceral cavity, and in *direct* communication only with the external medium. All the *Polyzoa* possess a nervous system, the characters and position of which are very well defined. It consists of a single ganglion (Fig. 9, *w*), placed between the oral and the anal apertures, and sending

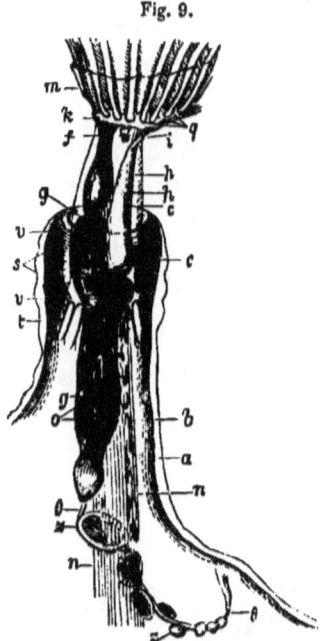

Fig. 9.

Fig. 9.—*Plumatella repens*, a single cell more magnified; *m*, calyx at the base of the ciliated tentacula borne by the disk or lophophore; *k*, gullet; *g g*, stomach; *h*, intestine; *i*, anus; *w*, nervous ganglion (after Allman).

off nerves in various directions. It has been affirmed that, in some *Polyzoa*, there is a more extended system of nerves by which the various zooids of the compound organism are placed in communication; but of that we want further evidence. In these animals no heart has been discovered as yet, the matters which result from digestion percolating through the walls of the intestine, and becoming mixed with the perivisceral fluid. One of the structural characters which I have mentioned is exceedingly important. As I have said, the intestine is not straight, but is bent upon itself (Fig. 9), and the direction of flexure is such that the nervous ganglion, which corresponds with those called "pedal" in *Lamellibranchiata*, is placed in the re-entering angle between the gullet and the rectum. In order to express this relation of the nervous system to the alimentary canal, the flexure of the latter has been called "neural"—the side of the body on which the principal ganglion is placed, and towards which the intestine is bent, being the "neural" side. Whatever our terminology, however, the great point is to remember that the structural relation which it expresses is constant throughout the *Polyzoa*.

IX. THE BRACHIOPODA.

Notwithstanding that these animals differ very much in external appearance from the *Polyzoa*, we shall find a singular fundamental resemblance of internal structure between the two classes. All known *Polyzoa* are compound animals, that is to say, the product of every ovum gives rise, by gemmation, to great assemblages of partially independent organisms, or zooids. The *Brachiopoda*, on the contrary, are all simple, the product of each ovum not giving rise to others by gemmation. All the *Brachiopoda* possess a bivalve shell—a shell composed of two, more or less horny, or calcified, pieces, which are capable of a certain range of motion on one another, and are very commonly articulated together by teeth and sockets. The proper body, which is small when compared with the size of the shell, has its dorsal integument produced into broad membranous expansions, which line the interior of the valves of the shell, and are called the lobes of the mantle or "pallium." The aperture of the mouth is situated in the middle line, between the pallial lobes, and, on each side of it, is a longer or shorter prolongation of the body, provided with ciliated tentacula. It is from the presence of these "arms" that the class has received its name. The tentaculate oral disk of a *Plumatella* is already horse-shoe shaped (Figs. 8 and 9); suppose each crus of the horse-shoe to be pulled out to a much greater length, and tentaculated "arms" would be produced, closely resembling those of the *Brachiopoda*.

The mouth leads into a gullet which is directed towards, or lies along, that side of the body from which one lobe of the

Fig. 10.—Lateral view of the viscera of *Waldheimia australis* (after Hancock). *a*, anterior layer of mantle; *b*, posterior layer; *c*, anterior walls of the body between the mantle lobes; *d*, arms; *p*, gullet; *q*, stomach, with cut biliary ducts of the left side; *r*, right hepatic mass; *s*, intestine ending cæcally between *j* and *k*; *v*, so-called "auricle" of the right "pseudo-heart," the left being almost wholly removed; *w*, pyriform vesicle fixed at the back of the stomach, and probably performing the function of a true heart; *z*, œsophageal ganglia.

mantle, the anterior, is continued; the gullet opens into a stomach, provided with a well-developed liver; and from the stomach, an intestine proceeds, which is directed towards, or along, that side of the body from which the other lobe of the mantle proceeds; and then either ends, blindly, in the middle line (Fig. 10), or else terminates in a distinct anus between the pallial lobes.

The principal ganglionic mass is situated behind and below the mouth, in the re-entering angle between the gullet and the rectum; in other words, the intestine has, as in the *Polyzoa*, a neural flexure (Fig. 10). In all *Brachiopoda* which have been carefully dissected a singular system of cavities and canals situated in the interior of the body, but in free communication with the surrounding medium, has been discovered. This, which I shall term the "atrial" system (from its close correspondence with the system of cavities, which has received the same name in the Ascidians), has been wrongly regarded as a part of the true vascular system, and the organs by which it is placed in communication with the exterior have been described as "hearts." There are sometimes two and sometimes four of these "pseudo-hearts," situated in that part of the body wall which helps to bound the pallial chamber. Each pseudo-heart is divided into a narrow, elongated, external portion (the so-called "ventricle"), which communicates, as Mr. Hancock has proved, by a small apical aperture with the pallial cavity; and a broad, funnel-shaped inner division (the so-called "auricle"), communicating on the one hand by a constricted neck with the so-called "ventricle," and, on the other, by a wide, patent mouth, with a chamber which occupies most of the cavity of the body proper, and sends more or less branched diverticula into the pallial lobes. These have been described as parts of the blood vascular system; and arterial trunks, which have no existence, have been imagined to connect the apices of the ventricles with vascular networks of a similarly mythical character, supposed to open into the branched diverticula.

In fact, as Mr. Hancock has so well shown in his splendid and exhaustive memoir, published in the *Philosophical Transactions* for 1857, the true vascular system is completely distinct from this remarkable series of "atrial" chambers and canals, the function of which would appear to be to convey away excretory matters and the products of the reproductive organs, which are developed in various parts of the walls of the atrial system.

The precise characters of the true vascular system of the *Brachiopoda* probably require still further elaboration than they have yet received; and the same may be said, notwithstanding the valuable contributions of F. Müller and of Lacaze Duthiers, of their development; but the shell, the pallial lobes, the intestine, and the nervous and the atrial systems, afford characters amply sufficient to define the class.

X. The Ascidioida.

These, like the *Brachiopoda*, are marine animals, and are very common all over the world; the more ordinary forms of them being always easily recognisable by the circumstance that their external integument is provided with two prominent, adjacent apertures, so that they look very much like double-necked jars (Fig. 11). At first sight you might hardly suspect the animal nature of one of these singular organisms, when freshly taken from the sea; but if you touch it, the stream of water which it squirts out of each aperture reveals the existence of a great contractile power within; and dissection proves that this power is exerted by an organism of a very considerable degree of complexity. Of the two apertures, the one which serves as a mouth is often—but not always—surrounded by a circlet of tentacles (Fig. 12, *c*). It invariably leads into an exceedingly dilated pharynx, the sides of which are, more or less extensively, perforated. The gullet comes off from the end of the pharynx, and then dilates into the stomach, from which an intestine, usually of considerable length, is continued to the anal aperture. The latter is almost always situated within a chamber which opens externally, by that second aperture upon the exterior of the test, to which I referred just now. Furthermore, in all Asci-

Fig. 11.—*Phallusia mentula*; *a*, oral; *b*, atrial aperture; *c*, base of attachment.

dians which I have examined, the first bend of the intestine takes place in such a manner that, if the intestine continued to preserve the same direction, it would end on the opposite side of the mouth to the nervous ganglion (Fig. 12); in other words, the nervous ganglion would not be situated in the re-entering angle between the gullet and the rectum, but on the opposite side of the gullet to that angle. Therefore, the flexure of the intestine is not neural, as in the *Polyzoa;* but as, on the contrary, the intestine is primarily bent towards the heart-side of the body, its flexure may be termed "hæmal." And this hæmal flexure of the intestine in the Ascidians thus constitutes an important element in the definition of the class.

In these animals there is an atrial system, the development of which is even more extraordinary than in the *Polyzoa.* The second aperture to which I have referred (*b*, Fig. 11 and *l*, Fig. 12) is continued into a large cavity, lined by a membrane, which is reflected, like a serous sac, on the viscera, and constitutes what is called the "third tunic," or "peritoneum." From the chamber which lies immediately within the second aperture (*k*, Fig. 12) it is reflected over both sides of the pharynx, extending, towards its dorsal part, very nearly as far as that structure which has been termed the "endostyle" (*m*, Fig. 12). It then passes from the sides of the

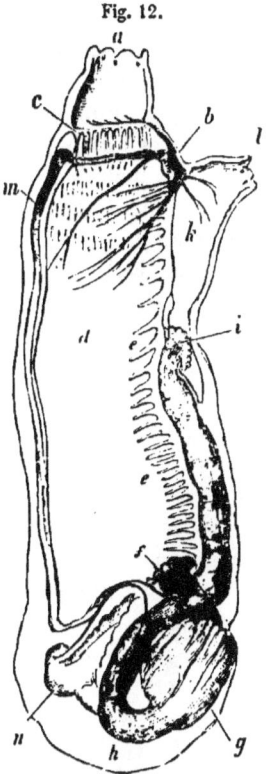

Fig. 12.—*Phallusia mentula;* the test removed, and hardly more of the animal drawn than would be seen in a longitudinal section. *a*, oral aperture; *l*, atrial aperture; *c*, circlet of tentacles; *d*, pharyngeal, or branchial, sac: the three rows of apertures in its upper part indicate, but do not represent, the pharyngo-atrial apertures; *e*, the languets: a series of tongue-shaped processes which project into the branchial sac; *f*, œsophageal opening; *g*, stomach; *h*, intestine performing its hæmal flexure; *i*, anus; *k*, atrium; *b*, ganglion; *m*, endostyle; *n*, heart.

pharynx to the body walls, on which the right and left lamellæ become continuous, so as to form the lining of the chamber (*k*), into which the second aperture (*l*) leads, or the "atrial chamber." Posteriorly, or at the opposite end of the atrial chamber to its aperture, its lining membrane (the "atrial tunic") is reflected to a greater or less extent over the intestine and circulatory organs, sometimes inclosing each of their parts in distinct plications (as in the genus *Phallusia*), sometimes merely passing over them, and limiting the blood sinus in which they are contained (as in *Clavelina*, &c.). Where the atrial tunic is reflected over the sides of the pharynx, the two enter into more or less close union, and the surfaces of contact become perforated by larger or smaller, more or less numerous, apertures. Thus the cavity of the pharynx acquires a free communication with that of the atrium; and, as the margins of the pharyngo-atrial apertures are fringed with cilia, working towards the interior of the body, a current is produced, which sets in at the oral aperture, and out by the atrial opening, and may be readily observed in a living Ascidian.

The Ascidians possess a distinct heart, but of a very simple construction, seeing that it is merely an incomplete muscular tube, open at each end, and devoid of valves. Functionally, it is not less remarkable than structurally; for, in the great majority of Ascidians, if not in all, it exhibits a regular alternation in the order of the peristaltic contractions of its muscular substance, which has no parallel in the Animal Kingdom. The result of this reversal in the direction of the contractions of the heart is a corresponding periodical reversal of the course of the circulation of the blood, so that the two ends of the heart are alternately arterial and venous.

The perforated pharynx performs the function of a branchial apparatus, the blood contained in its sieve-like walls being subjected to the action of constant currents of aërated water. All Ascidians possess a single nervous ganglion placed upon one side of the oral aperture (*b*, Fig. 12), and, in all known genera but *Appendicularia*, it is situated between the oral and atrial apertures, and, indeed, between the oral and anal apertures; for, in all genera but that mentioned, the intestine, after it has made its hæmal bend, curves down towards the neural side of the body, and opens into the atrium on that side of the body, and behind the nervous ganglion.

The outer integument of the Ascidians secretes upon its surface, not a calcareous shell, but a case or "test," which may vary in consistence from jelly to hard leather or horn. And it is not one of the least remarkable characteristics of the group that this test is rendered solid, by impregnation with a substance identical in all respects with that "cellulose" which is the proximate principle of woody fibre, and forms the chief part of the skeleton of plants. Before the discoveries of late years had made us familiar with the production of vegetable proximate principles by the metamorphosis of animal tissues, this circumstance was justly regarded as one of the most remarkable facts of comparative physiology.

The last common and distinctive peculiarity of the Ascidians which I have to mention, is one which acquires importance only from its constancy. The middle of the hæmal wall of the pharynx, from near the oral to the œsophageal end, in all these animals, is pushed out into a longitudinal fold, the bottom of which projects into a blood sinus, and has a much thickened epithelial lining. Viewed from one side, the bottom of the fold consequently appears like a hollow rod, and has been termed the "endostyle" (m, Fig. 12). The functions of this structure are unknown, but it has been noticed in all genera of Ascidians hitherto examined.

XI. THE LAMELLIBRANCHIATA.

This group, comprising those creatures which we know as mussels, cockles, and scallops, and all the fabricators of what are commonly known as bivalve shells (except the *Brachiopoda*), presents an important advance in organization. In all these animals, the body, as is exemplified by the diagram (Fig. 13) of a freshwater mussel (*Anodon*), is included within a mantle or "pallium," which is formed by a prolongation of the dorsal integument,—a structure in principle quite similar to that which we met with in the *Brachiopoda*. But there is this essential difference between the two,—that whereas, in the *Brachiopoda*, the mantle lobes corresponded with the anterior and posterior regions of the body, in the *Lamellibranchiata* they answer to the right and left halves of the body. The intestine, which always terminates by a definite anus between the mantle lobes, at the posterior end of the body, has its first flexure neural. There is always a well-developed heart, which is much more complex than that of the Ascidians or

Fig. 13.

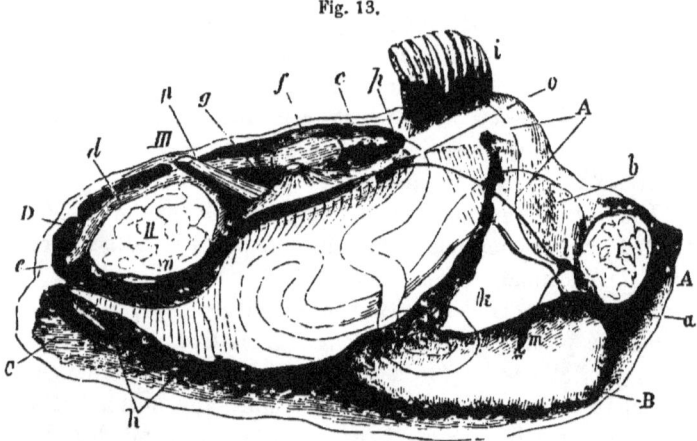

Fig. 13.—Sectional diagram of a freshwater mussel (*Anodon*). *A A*, mantle, the right lobe of which is cut away; *B*, foot; *C*, branchial chamber of the mantle cavity; *D*, anal chamber; *I*, anterior adductor muscle; *II*, posterior adductor muscle; *a*, mouth; *b*, stomach; *c*, intestine, the turns of which are supposed to be seen through the side walls of the foot; *d*, rectum; *e*, anus; *f*, ventricle; *g*, auricle; *h*, gills, except *i*, right external gill, largely cut away and turned back; *k*, labial palpi; *l*, cerebral; *m*, pedal; *n*, parieto-splanchnic ganglia; *o*, aperture of the organ of Bojanus; *p*, pericardium.

Brachiopods, being divided into distinct auricular and ventricular chambers. Commonly, there are two auricles and one ventricle, as in *Anodon*; but in other *Lamellibranchiata*, such as the oyster, there is a single auricle and a single ventricle, and, in some exceptional cases, there are two auricles and two ventricles, forming two distinct hearts. Distinct respiratory organs, which usually have the form of lamellæ (as the name of the class implies), are found in all *Lamellibranchiata*, and are situated upon each side of the body, in a chamber which extends between the foot and the mantle lobes in front, and between the mantle lobes posteriorly (Fig. 14). The branchial organs may consist of distinct filaments, or of plates composed of tubular rods supporting a network of blood-vessels, and covered with cilia, by the action of which they are constantly bathed by currents of water.

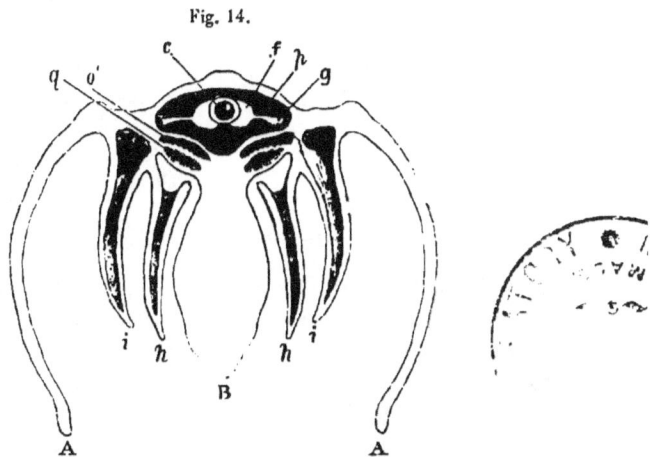

Fig. 14.—*Anodon*, vertical and transverse section of the body through the heart; *f*, ventricle; *g*, auricles; *c*, rectum; *p*, pericardium; *h*, inner, *i*, outer gill; *o'q*, organ of Bojanus; B, foot; A A, mantle lobes.

The nervous system presents a no less distinct advance than the other organs. All Lamellibranchs possess at least three pairs of principal ganglia—a cerebral pair at the sides of the mouth, a pedal pair in the foot, and a third pair on the under surface of the posterior adductor muscle, which are commonly called "branchial," but which, as they supply not only branchial, but visceral and pallial filaments, may more properly be termed "parieto-splanchnic." Three sets of commissural filaments connect the cerebral ganglia with one another, with the pedal, and with the parieto-splanchnic ganglia. The inter-cerebral commissures surround the mouth, and the other two pairs of cords extend respectively from the cerebral to the pedal, and from the cerebral to the parieto-splanchnic ganglia.

Finally, there is always, in these animals, an external shell, which is formed as an excretion from the surface of the lobes of the mantle, and is composed of layers of animal matter hardened by deposit of carbonate of lime, which may or may not take a definite form, and so give rise to "prismatic" and "nacreous" substance. As the lobes are right and left, so the valves of the shell are right and left, and differ altogether from the valves of the shell of the *Brachiopoda*, which are anterior and posterior. The valves of the shell can be brought together by adductor muscles. Of these one (Fig. 13, *II*) always exists, posteriorly, on the neural side of the intestine. A second (Fig. 13, *I*) is commonly found anteriorly to the mouth, on the hæmal side of the intestine.

XII. THE BRANCHIOGASTEROPODA.

The hiatus between the present class and that just defined is considerable, though not quite so well marked as that between the Ascidians and the *Lamellibranchiata*. This group, which contains the whelks, periwinkles, sea-slugs, or the *Heteropoda, Pectinibranchiata, Cyclobranchiata, Tectibranchiata, Inferobranchiata*, and *Nudibranchiata*, of Cuvier, consists of animals which, like the Lamellibranchs, possess (in their young state, at any rate) a mantle; a foot, which is the chief organ of locomotion; and three principal pairs of ganglia—cerebral, pedal, and parieto-splanchnic. When they are provided with a heart, which is usually the case, it is divided into auricular and ventricular chambers; but the mantle, instead of being divided into two lateral lobes, is continuous round the body, and when it secretes a shell from its surface, that shell is either in a single piece, or the pieces are repeated from before backwards, and not on each side of the median line. The shell of a Branchiogasteropod may, therefore, be univalve,—composed of a single conical piece, straight or coiled; or it may be multivalve—formed of a number of segments succeeding one another antero-posteriorly; but it is never bivalve.

Sometimes a shelly, horny, or fibrous secretion takes place from the foot, giving rise to a structure resembling the byssus of some Lamellibranchs; it is the so-called "*operculum*," which serves to protect the animal when retracted into its shell: but as the operculum is developed from the foot and not from the mantle, it can obviously have no homology with the valves of either a Brachiopod or a Lamellibranch. The *Branchiogasteropoda* (Fig. 15) commonly possess a distinct head, provided with

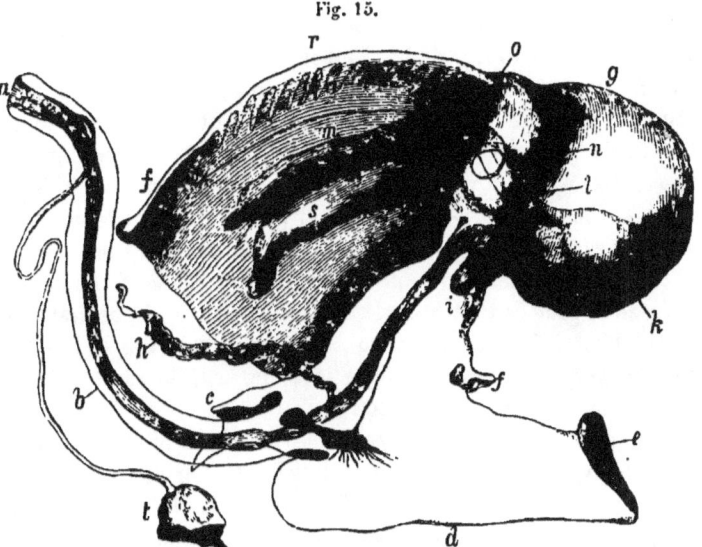

Fig. 15.

Fig. 15.—Section of a female whelk (*Buccinum*). The organs marked *t* and *h* are removed from their proper places; the others are seen *in situ*. *a*, mouth; *b*, gullet; *c*, head; *d*, foot; *e*, operculum; *f*, free part of the mantle; *g*, that part which invests the visceral mass lodged within the shell; *h*, a gland of unknown function connected with the gullet; *i*, crop; *k*, stomach; *l*, intestine; *m*, rectum; *n*, heart; *o*, aperture of the renal organ; *r*, mucous gland developed from the wall of the mantle cavity; *s*, oviduct; *t*, salivary gland. The arrows indicate the position of the branchiæ. The cerebral, pedal, and parieto-splanchnic ganglia closely surround the gullet, and the latter send off a long ganglionated cord towards the heart and branchiæ.

a pair of tentacles and a single pair of eyes, both of which are supplied with nerves from the cerebral ganglia. Cephalic eyes of this kind are not known in the *Lamellibranchiata*. But the characters which most definitely distinguish the *Branchiogasteropoda* are to be found in the alimentary canal. The cavity of the mouth is invariably provided with an organ which is usually, though not very properly, called the tongue, and which might more appropriately be denominated the "odontophore." It consists essentially of a cartilaginous cushion, supporting, as on a pulley, an elastic strap, which bears a long series of transversely disposed teeth. The ends of the strap are connected with muscles attached to the upper and lower surface of the hinder extremities of the cartilaginous cushions; and these muscles, by their alternate contractions, cause the toothed strap to work, backwards and forwards, over the end of the pulley formed by its anterior end. The strap consequently acts, after the fashion of a chain-saw, upon any substance to which it is applied, and the resulting wear and tear of its anterior teeth are made good by the incessant development of new teeth in the secreting sac in which the hinder end of the strap is lodged. Besides the chain-saw-like motion of the strap, the odontophore may be capable of a licking or scraping action as a whole.

The other peculiarity of the alimentary canal of the *Branchiogasteropoda* is that it is always bent upon itself, at first, not to the neural, but to the hæmal, or heart side of the body—the rectum very commonly opening into the mantle cavity, above the cephalic portion of the body.

In most *Branchiogasteropoda* the foot is a broad, flat, muscular body, without any distinct division of parts; but in some forms, such as the *Heteropoda* of Cuvier, it is divided into three very well-marked portions—an anterior, a middle, and a posterior, which are termed respectively the *propodium, mesopodium,* and *metapodium;* while the *Aplysiæ,* in which the foot proper has the ordinary composition, exhibit processes from the lateral and upper surfaces of that organ, having the form of great muscular lobes, which serve as a sort of aquatic wings to some species, and are termed *epipodia.*

The *Branchiogasteropoda* are such of the *Gasteropoda* of Cuvier as breathe water either by means of the thin wall of the mantle cavity (*Atlanta, e.g.*), or by special pallial branchiæ (*Pectinibranchiata, Tectibranchiata,* &c.), or by certain parts of the integument of the body (*Nudibranchiata*) more or less specially modified.

XIII. THE PULMOGASTEROPODA.

These are the Pulmonate *Gasteropoda* of Cuvier, or the snails and slugs, which agree with the *Branchiogasteropoda* in the general characters of their body, mantle, nervous and respiratory systems, and in possessing an odontophore; but differ from them, not only in breathing air by means of the thin lining of the pallial chamber, but, as I believe, by the direction of the flexure of their intestine. A careful dissection of a common snail, for example (Fig. 16), will prove that, though the anus is

Fig. 16.

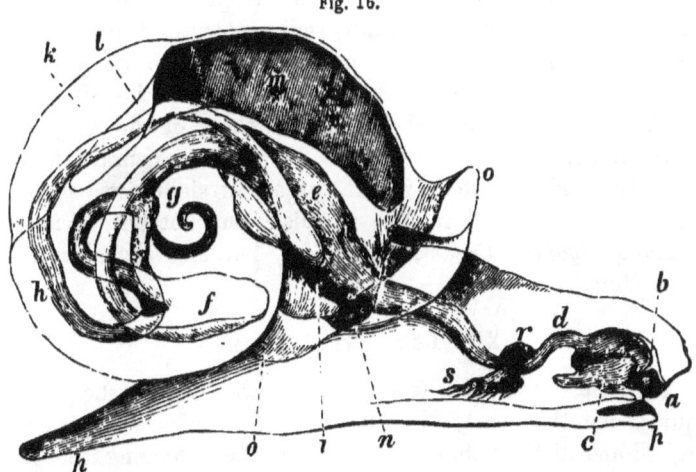

Fig. 16.—Diagram exhibiting the disposition of the intestine, nervous system, &c., in a common snail (*Helix*). *a*, mouth; *b*, tooth; *c*, odontophore; *d*, gullet; *e*, its dilatation into a sort of crop; *f*, stomach; *g*, coiled termination of the visceral mass; the letter is also close to the commencement of the intestine, which will be seen to lie *under* the œsophagus, and not over it as in the whelk; *h*, rectum; *i*, anus; *k*, renal sac; *l*, heart; *m*, lung, or modified pallial chamber; *n*, its external aperture; *o*, thick edge of the mantle united with the sides of the body; *p*, foot; *r*, cerebral, pedal, and parieto-splanchnic ganglia aggregated round the gullet.

situated in the same way as in the *Branchiogasteropoda*, on the dorsal or hæmal side of the body, the primary bend of the intestine is not to the hæmal, but to the neural side, the eventual termination of the intestine on the hæmal side being the result of a second change in its direction.

How far this neural flexure of the intestine really prevails among the Pulmogasteropods is a question which must be decided by more extensive investigations than I have as yet been enabled to carry out.

XIV. THE PTEROPODA.

The members of this class are small, or even minute, molluscs; all marine in habit, and for the most part pelagic, or swimmers at the surface of deep seas. Like the two preceding groups, they possess three principal pairs of ganglia; an odontophore; a mantle, which is not divided into two lobes, and which secretes a univalve shell, if any. But the propodium, mesopodium, and metapodium are usually rudimentary, and locomotion is almost wholly effected by the epipodia, which are enormously developed, and, in most of the genera, perform the office of aquatic wings still more efficiently than those of the *Aplysiæ*. Furthermore, the intestine is flexed towards the neural side of the body; and the head, with the organs of sight, are usually quite rudimentary. I include in this group not only *Criseis*, *Cleodora*, *Hyalæa*, *Pneumodermon*, &c., but also the aberrant genus *Dentalium*.*

XV. THE CEPHALOPODA.

This class comprises the Poulpes, the Cuttle-fishes, the Squids, and the pearly Nautilus; and it is definable by most marked and distinct characters from all the preceding, though it resembles them in fundamental characters. Thus, the mantle is related to the body, as in *Pteropoda* and *Gasteropoda*; when an external shell exists it is composed of a single piece; and the Cephalopods have an odontophore constructed upon just the same principle as that of the other classes. The nervous system, the foot, and the epipodia exhibit the same primary relations as in these groups, and there is a distinct head, with ordinarily well-developed optic and olfactory organs. That which essentially characterises the *Cephalopoda*, in fact, is simply the manner in which, in the course of development, the margins

---

* *Dentalium* resembles the *Pteropoda* in its rudimentary head, the neural flexure of its intestine, its epipodial lobes, and the character of its larva.

of the foot proper and the epipodia become modified and change their relations. The margins of the foot are produced into more or less numerous tentacular appendages, often provided with singularly constructed suckers, or *acetabula ;* and the antero-

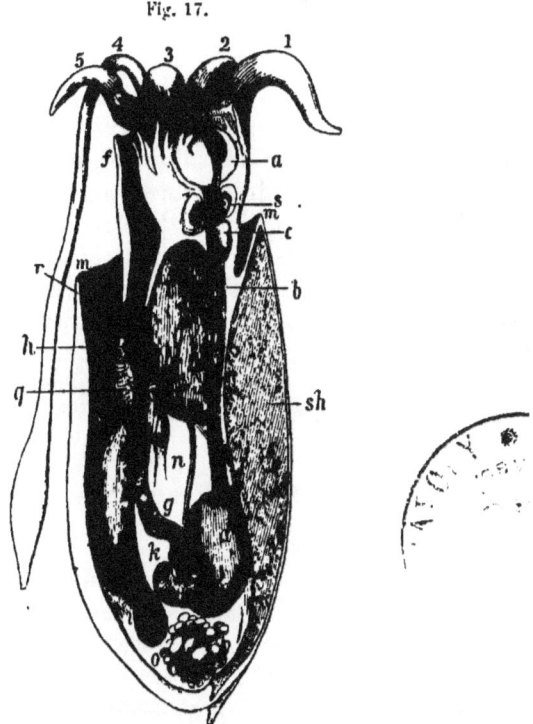

Fig. 17.—Diagrammatic section of a female Cephalopod (*Sepia officinalis*). *a,* Buccal mass surrounded by the lips, and showing the horny jaws and tongue; *b,* œsophagus; *c,* salivary gland; *d,* stomach; *e,* pyloric cæcum; *f,* the funnel; *g,* the intestine; *h,* the anus; *i,* the ink-bag; *k,* the place of the systemic heart; *l,* the liver; *n,* the hepatic duct of the left side; *o,* the ovary; *p,* the oviduct; *q,* one of the apertures by which the atrial system of water-chambers is placed in communication with the exterior; *r,* one of the branchiæ; *s,* the principal ganglia aggregated round the œsophagus; *m,* the mantle; *sh,* the internal shell, or cuttle-bone. 1, 2, 3, 4, 5, the produced and modified margins of the foot, constituting the so-called "arms" of the *Sepia*.

lateral parts of each side of the foot extend forwards beyond the head, uniting with it and with one another; so that, at length, the mouth, from having been situated, as usual, above the anterior margin of the foot, comes to be placed in the midst of it. The two epipodia, on the other hand, unite posteriorly above the foot, and where they coalesce, give rise either to a folded muscular expansion, the edges of which are simply in apposition, as in *Nautilus;* or to an elongated flexible tube, the apex of which projects beyond the margin of the mantle (Fig. 17, *f*), and is called the funnel or *infundibulum*, as in the dibranchiate *Cephalopoda*.

The *Cephalopoda* present a vast number of the most interesting features, to which it would be necessary to devote much attention if we were studying all the organic peculiarities manifested by the class; but it is in the characters of foot and of the epipodium that the definition of the class must be chiefly sought. In addition, the flexure of the intestine is, in all Cephalopods, neural; and the mouth is always provided with a horny or more or less calcified beak, like that of a parrot, composed of two curved pieces, which move in the median antero-posterior plane of the body; and one of which, that on the neural side, is always longer than the other.

XVI. THE ECHINODERMATA.

The star-fishes, sea-urchins, sea-cucumbers, trepangs, and feather-stars—known technically as *Asteridea, Echinidea, Holothuridea, Ophiuridea, Crinoidea*, &c., are marine animals which differ vastly in external appearance, though they all, in the adult state, present a more or less definitely radiate arrangement of some parts of their organization.

That which most remarkably distinguishes the *Echinodermata* is the nature of the embryo, and the strange character of the process by which the adult form is originated by a secondary development within that embryo.

In the great majority[*] of the *Echinodermata*, the development of which has been examined, the impregnated egg gives rise to a free-swimming, ovoid, ciliated embryo, the cilia of

---

[*] In *Ophiolepis squamata* and *Echinaster sepositus*, the larva appears to attain only a very imperfect state of development before the appearance of the echinoderm body; and careful re-examination is required to decide how far the larvæ of these animals are truly bilateralllly symmetrical.

which soon become restricted to, and, at the same time, largely developed upon, one, two, or more bands, which are disposed either transversely, or more or less obliquely to the longitudinal axis of the body, but which are, in any case, bilaterally symmetrical (Fig. 18).

Fig. 18.

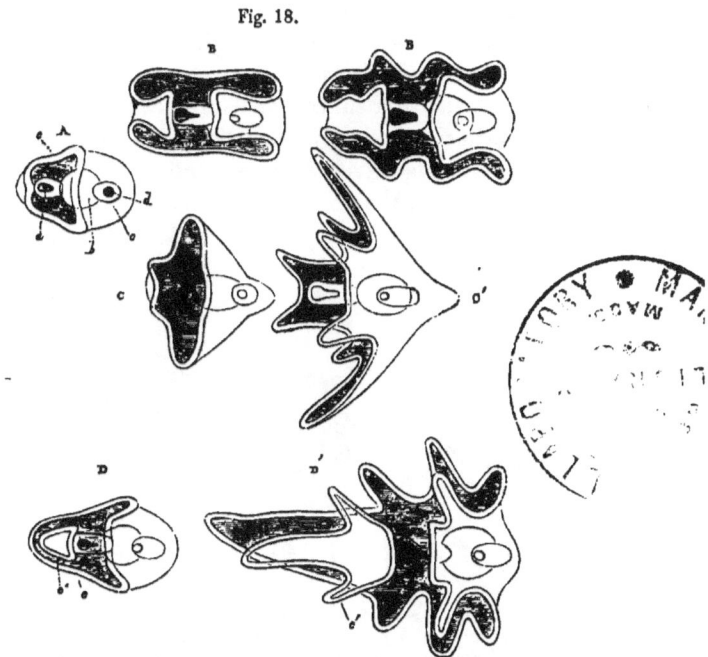

Fig. 18.—Diagram exhibiting the general plan of the development of the *Echinoderms* (after Müller).—A. Common form whence the Holothurid (B, B′) and Ophiurid or Echinid (C, C′) larvæ are derived. D, D′. Younger and more advanced stages of the Asterid (*Bipinnaria*) larvæ. *a*. Mouth. *b*. Stomach. *c*. Intestine. *d*. Anus. *e*. Ciliated band. *e′*. Second or anterior ciliated circlet.

The parts of the body which carry the ciliated band, or bands, often become developed into processes, which correspond upon each side of the body, and thus render its bilateral symmetry more marked (Fig. 18, C′, D′). And, in the larvæ of some *Echinidea* and *Ophiuridea*, other bilaterally symmetrical processes are developed from parts of the body which do not lie in the course of the ciliated bands.

The larvæ of *Asteridea* and *Holothuridea* are devoid of any continuous skeleton, but those of *Ophiuridea* and *Echinidea* possess a very remarkable bilaterally symmetrical, continuous, calcareous skeleton, which extends into, and supports the processes of the body (Fig. 21).

A distinctly defined alimentary canal early makes its appearance in these Echinoderm larvæ. It is divided into a well-marked oral and œsophageal portion, a globular stomach, and a short intestine terminating in an anal aperture (Figs. 18 and 19). All the parts of the alimentary canal are disposed in a longitudinal and vertical plane, dividing the larval body into two symmetrical halves; but the œsophageal and intestinal portions are so disposed as to make an angle, open towards the ventral side, with one another. No nervous, or other organs, besides those indicated, have as yet been discovered in these larvæ.

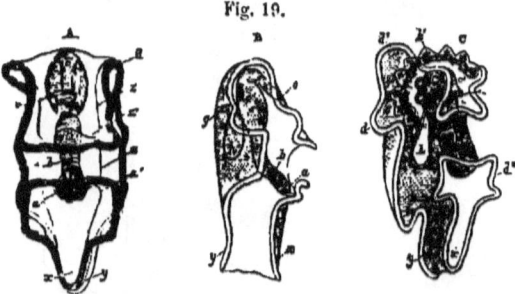

Fig. 19.

Fig. 19.—A young *Asterid* larva (after Müller).—A, Ventral. B, Lateral view of the larva. C, Commencing rudiment of the starfish. *a*, Mouth. *b*, Œsophagus. *c*, Stomach. *c'*, Intestine. *o*, Anus. *x*, Anterior, and *y*, principal ciliated band. *h*, Cæcal diverticulum, forming the rudiment of the ambulacral vascular system, and opening externally by the pore, *g*. *k'*, Perisoma of the starfish.

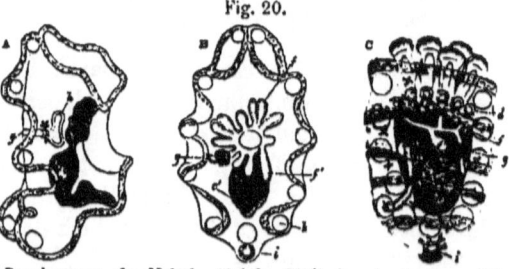

Fig. 20.

Fig. 20.—Development of a *Holothurid* (after Müller).—A, Early condition of larva. B, C. Later stages. *f*, *g*, *h*, the ambulacral vascular system.

THE ECHINODERMATA.

Fig. 21.

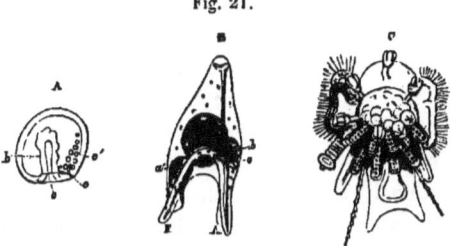

Fig. 21.—Development of an *Echinid* larva (after Müller).—A, earliest, and B, later condition of larva. C. The Echinid imago developed within and nearly obliterating the larva.

After swimming about in this condition for a while, the larva begins to show the first signs of those changes by which it is converted into the adult Echinoderm. An involution of the integument takes place upon one side of the dorsal region of the body, so as to give rise to a cæcal tube, which gradually elongates inwards, and eventually reaches a mass of formative matter, or blastema, aggregated upon one side of the stomach. Within this, the end of the tube becomes converted into a circular vessel, from which trunks pass off, radially, through the enlarging blastema. The latter, gradually expanding, gives rise in the *Echinidea*, the *Asteridea*, the *Ophiuridea*, and the *Crinoidea*, to the body-wall of the adult; the larval body and skeleton (when the latter exists), with more or less of the primitive intestine, being either cast off as a whole, or disappearing, or becoming incorporated with the secondary development, while a new mouth is developed in the centre of the ring formed by the circular vessel. The vessels which radiate from the latter give off diverticula to communicate with the cavities of numerous processes of the body—the so-called feet—which are the chief locomotive organs of the adult. The radiating and circular vessels, with all their appendages, constitute what is known as the "ambulacral system;" and, in Asterids and Echinids, this remarkable system of vessels remains in communication with the exterior of the body by canals, connected with perforated portions of the external skeleton—the so-called "madreporic canals" and "tubercles." In Ophiurids the persistence of any such communication of the ambulacral system with the exterior is doubtful, and still more so in Crinoids. In Holothurids no such communication obtains, the madreporic canals and their tubercles depending freely from the circular canal into the perivisceral cavity.

Whether the larva possessed a skeleton or not, the adult Echinoderm presents a calcareous framework which is developed quite independently of that of the larva. This skeleton may be composed of mere detached spicula, or plates, as in the Holothurids; or of definitely disposed ossicula, or regular plates, as in other Echinoderms. In the latter case its parts are always disposed with a certain reference to the disposition of the ambulacral system, and hence have a more or less distinctly radiate arrangement. It might be expected, in fact, that the arrangement of the organs of support should follow more or less closely that of the chief organs of movement of the adult Echinoderm, and it is not surprising to find the nervous system similarly related. It is, in all adult Echinoderms, a ring-like, or polygonal, ganglionated cord, situated superficially to that part of the ambulacral system which surrounds the mouth, and sending prolongations parallel with, and superficial to, the radiating ambulacral trunks.

The reproductive organs of the Echinoderms, which usually open upon, or between, parts of the radially disposed skeleton, commonly partake of the radial symmetry of that skeleton; but they have no such radial symmetry in the *Holothuridea.*

The alimentary canal of the adult Echinoderm is still less dependent upon the skeleton, and only in one group, the *Asteridea*, exhibits anything approaching a radiate disposition. Where skeletal elements are developed around the mouth or gullet, however, they have a radial disposition; as, *e. g.*, the parts of the so-called "lantern of Aristotle."

The vascular system which exists in many, if not all, adult Echinoderms, but the true nature of which is by no means understood at present, is closely related both to the alimentary and to the ambulacral systems, and partakes of the disposition of both.

No Echinoderm whatsoever has its organs, internal or external, disposed with that absolute and perfect radial symmetry which is exhibited by a *Medusa*, the tendency towards that kind of symmetry being always disturbed, either by the disposition of the alimentary canal, or by that of some part of the ambulacral apparatus. Very often, as in the Spatangoid sea-urchins, and in many *Holothuridea*, the ambulacral and nervous systems alone exhibit traces of a radial arrangement; and in the larval state, as we have seen, radial symmetry is totally absent, the young Echinoderm exhibiting as complete a bilateral symmetry as Annelids, or Insects.

XVII. The Scolecida.

Nothing can be more definite, it appears to me, than the class *Echinodermata*, the leading characteristics of which have just been enumerated; but it is a very difficult matter to say whether the seven groups, some of considerable extent, which are massed under the present head, are rightly associated into one class, or should be divided into several. The seven groups to which I refer are the *Rotifera* (or Wheel-animalcules), the *Turbellaria*, the *Trematoda* (or flukes), the *Tæniada* (or tapeworms), the *Nematoidea* (or threadworms), the *Acanthocephala*, and the *Gordiacea*. Of these, five are composed of animals parasitic upon others; and exhibiting the anomalies of structure and of development which might be expected from creatures living under such exceptional conditions.

There is one peculiarity of organic structure which the first four of these groups certainly have in common; they all present what is termed the " water-vascular system,"—a remarkable set of vessels which communicate with the exterior by means of one, or more, apertures situated upon the surface of the body, and branch out, more or less extensively, into its substance.

In the *Rotifera* the external aperture of the water-vascular system is single, and situated at the hinder end of the body; it communicates with a large, rhythmically contractile, sac, whence two trunks proceed, which usually give off short lateral branches, and terminate in the ciliated "trochal disk" of the Rotifer, in the middle of which its mouth is placed. Both the lateral offshoots and the terminal branches contain vibratile cilia. The Trematode and Tænioid worms have a similar, but usually much more ramified apparatus; and it is interesting to observe that, in these animals, as in the *Aspidogaster conchicola* (Fig. 22), the water-vascular system becomes divided into two

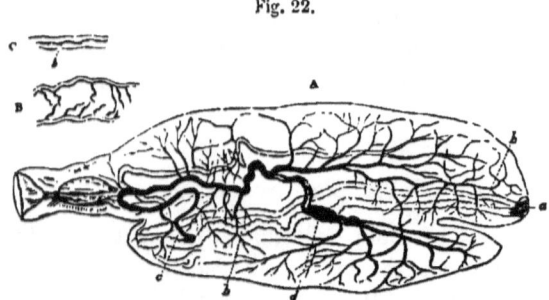

Fig. 22.

Fig. 22.—A, Water-vascular system of *Aspidogaster conchicola*; *a*, terminal pore; *b*, lateral contractile vessels; *c*, lateral ciliated trunks, that of the left side shaded; *d*, dilatation of this trunk; B, one of the larger, and C, one of the smaller, ciliated vessels.

distinct portions, one with contractile and non-ciliated walls, the other with non-contractile and ciliated walls. In some *Turbellaria* the apertures of the water-vascular apparatus are multiple; while it would seem that in others, as the *Nemertidæ*, the apparatus becomes shut externally in the adult state, and consists mainly, if not exclusively, of contractile vessels. The difficulties of observation are here, however, very great, and I would be understood to express this opinion with all due caution.

In none of these animals has any other set of vessels than those which appertain to the water-vascular system (if I am right in my view of the vessels of the *Nemertidæ*) been observed, nor has any trace of a true heart been noticed. The nervous system consists of one, or two, closely approximated ganglia.

This sum of common characters appears to me to demand the union of the *Rotifera*, *Turbellaria*, *Trematoda*, and *Tæniada* into one great assemblage. Ought the Nematoid worms to be grouped with them? If the system of canals, in some cases contractile, which open externally near the anterior part of the body (Fig. 23), and were originally observed by Von Siebold, and since by myself and others, are to be regarded as homologous with the water-vessels of the *Trematoda*, this question must, I think, be answered in the affirmative. It is almost the only system of organs in the *Nematoidea* which gives us a definite zoological criterion, the condition of the nervous system in these animals being still, notwithstanding the many inquiries which have been made into the subject, a matter of great doubt.

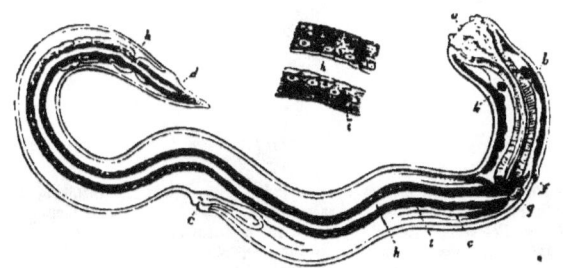

Fig. 23.—*Oxyuris.*—*a.* Mouth. *b.* Pharynx. *c.* Commencement of intestine, and *d* its termination; the intermediate portion is not figured. *e.* Genital aperture. *f.* Opening of vessels. *g.* Their receptacle. *h.* One of the vessels. *i.* Cellular matter enveloping them. A portion of one of the contractile vessels is represented above, more highly magnified.

In habit and feature, the *Gordiacea*, filiform parasites which inhabit the bodies of insects, and leave their hosts only to breed, resemble the *Nematoidea* so much that there can be no doubt that their systematic place must be close to that of the latter.

The structure of the *Acanthocephala*, comprising the formidable *Echinorhynchus* (Fig. 24) and its allies, is pretty clearly made out. They are vermiform parasites, like the *Tæniada*, devoid of any mouth or alimentary canal, but provided with a proboscis armed with recurved hooks. The proboscis is supported within by a sort of rod-like handle, whence a cord is continued, to which the reproductive organs are attached. A single ganglion is seated in the "handle" of the proboscis. Immediately beneath the integument lies a series of reticulated canals containing a clear fluid, and it is difficult to see with what these can correspond if not with some modification of the water-vascular system.*

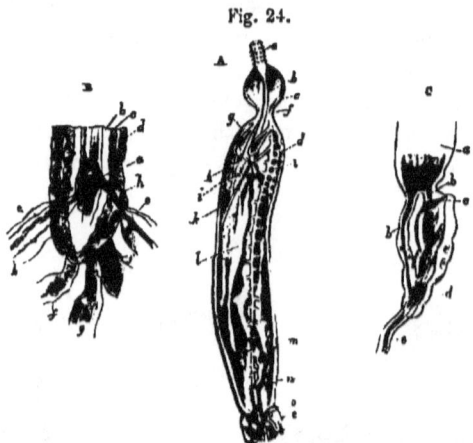

Fig. 24.

Fig. 24.—The *Echinorhynchus* of the Flounder.—A. Diagram exhibiting the relative position of the organs. *a*. Proboscis. *b*. Its stem. *c*. Anterior enlargement. *d*. Body. *e*. Posterior "funnel." *f*. Neck. *g*. Meniscus. *h*. Superior oblique tubular bands. *h*. Inferior muscles of the proboscis. *l, m*. Genitalia. *o*. Penis, or vulva. B. Lower extremity of the stem of the proboscis. *a*. Ganglion. *b*. Interspace. *d*. Outer coat. *c*. Inner wall. *e*. Tubular band, with the nerve *h*. *f*. Muscular bands. *g*. Suspensorium of the genitalia. C. Part of the female genitalia. *a*. Ovary. *b b*. Ducts leading from ovary to uterus (spermiducts?). *c*. Open mouth of oviduct. *d e*. Uterus and vagina.

XVIII. THE ANNELIDA.

This is a class of large extent, containing the leech, the earthworm, the *Sipunculus*, the lobworm, the seamouse and *Polynoë* (Fig. 25), the *Serpula*, and the *Spirorbis*.

All the members of this class possess a nervous system, which consists of a longitudinal series of ganglia, situated along one side of the body, and is traversed anteriorly by the œsophagus; the præ-œsophageal, or so-called " cerebral," ganglia

---

* The investigations of Leuckart, while they demonstrate still more clearly the close affinity which exists between the *Acanthocephala* and the *Tæniada*—by proving the adult worm to arise by secondary growth within a hooked embryo, in the former case as in the latter—leave some doubt upon the nature of the reticulated canals. According to Leuckart, they are the remains of the cavity which primitively lies between the wall of the embryo and the contained rudiment of the adult Acanthocephalan body.

being connected by lateral commissural cords with the post-œsophageal ganglia.

In many of these animals the body is divided into segments, each of which corresponds with a single pair of ganglia of the chain, and each of these segments may be provided with a pair of lateral appendages; but the appendages are never articulated; and are never so modified, as to be converted into masticatory organs, in the cephalic region of the body.

No Annelid ever possesses a heart comparable to the heart of a Crustacean, or Insect; but a system of vessels, with more or less extensively contractile walls, containing a clear fluid, usually red or green in colour, and, in some rare cases only, corpusculated, is very generally developed, and sends prolongations into the respiratory organs, where such exist. This has been termed the "pseudo-hæmal" system; and I have

Fig. 25.

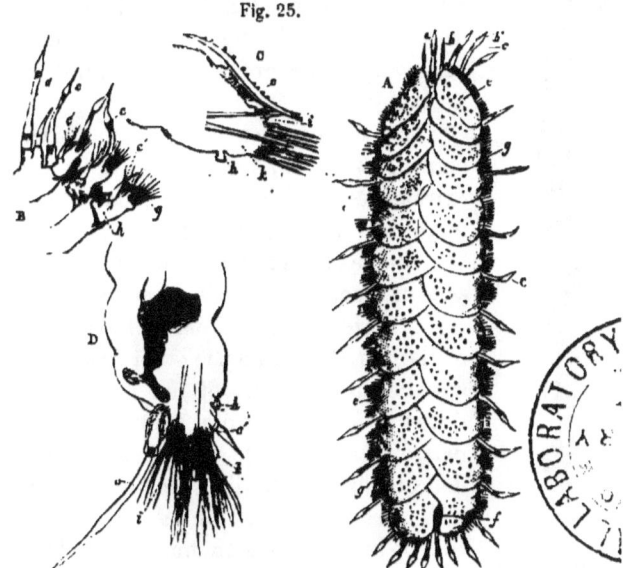

Fig. 25.—*Polynoë squamata.*
A. Viewed from above and enlarged. *a, b.* Feelers. *c.* Cirri. *e.* Elytra. *f.* Space left between the two posterior elytra. *g.* Setæ and fimbriæ of the elytra.
B. Posterior extremity, inferior view, the appendages of the left side being omitted. *h.* Inferior tubercle.
C. Section of half a somite with elytron. *i.* Notopodium. *k.* Neuropodium.
D. Section of half a somite with cirrus.

thought it probable that these "pseudo-hæmal" vessels are extreme modifications of organs homologous with the water-vessels of the *Scolecida*. As M. de Quatrefages has clearly shown, it is the perivisceral cavity with its contents that, in these animals, answers to the true blood-system of the Crustacea and Insects.

The embryos of Annelids are very generally ciliated, and vibratile cilia are commonly, if not universally, developed in some part or other of their organization. In both these respects they present a most marked contrast to the succeeding classes.

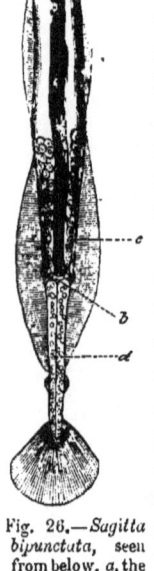

Fig. 26.—*Sagitta bipunctata*, seen from below. *a*, the head with its eyes and appendages; *b*, the anus; *c*, the ovary; *d*, the testicular chambers.

XIX. THE CHÆTOGNATHA.

There is a very aberrant marine genus, *Sagitta*, the nearest affinities of which appear to be with the *Annelida*, but which is so unlike them and every other group as to require a class for itself.

The *Sagittæ* are elongated and transparent animals, with rounded heads and tapering caudal extremities, which do not usually attain much more than an inch in length.

The head is provided with several, usually six, sets of strong bilaterally symmetrical oral setæ, two of which, long and claw-like, lie at the sides of the mouth; while the other four sets are short, and lie on that part of the snout which is produced in front of the oral aperture. The posterior part of the body is fringed on each side by a delicate striated fin-like membrane, which seems to be an expansion of the cuticle. In some species the body is beset with fine setæ. The intestine is a simple, straight tube, extending from the mouth to the anus; the latter opens on the ventral surface, just in front of the hinder extremity. A single oval ganglion lies in the abdomen, and sends, forwards and backwards, two pairs of lateral cords. The anterior cords unite in front of and above the mouth, into a hexagonal ganglion. This gives off two branches which dilate at their extremities into the spheroidal ganglia, on which the darkly pigmented imperfect eyes rest. The ovaries, saccular organs, lie on each side of the intestine and open on either side of the vent; *receptacula seminis* are present. Behind the anus, the cavity of the tapering caudal part of the body is partitioned into two compartments; on the lateral parietes of these, cellular masses are developed which become detached, and floating freely in the compartment, develope into spermatozoa. These escape by spout-like lateral ducts, the dilated bases of which perform the part of *vesiculæ seminales*. The embryos are not ciliated, and undergo no metamorphosis.

XX. THE CRUSTACEA.

In this class (Fig. 27), the body is distinguishable into a variable number of "somites," or definite segments, each of which may be, and some of which always are, provided with a single pair of articulated appendages. The latter proposition is true of all existing *Crustacea*: whether it also held good of the long extinct *Trilobites*, is a question which we have no means of deciding. In most *Crustacea*, and, probably in all, one or more pairs of appendages are so modified as to subserve manducation. A pair of ganglia is primitively developed in each somite, and the gullet passes between two successive pairs of ganglia, as in the *Annelida*.

No trace of a water-vascular system, nor of any vascular system similar to that of the *Annelida*, is to be found in any Crustacean. All *Crustacea* which possess definite respiratory organs have branchiæ, or outward processes of the wall of the body, adapted for respiring air by means of water; the terrestrial *Isopoda*, some of which exhibit a curious rudimentary representation of a tracheal system, forming no real exception to this rule. When they are provided with a circulatory organ, it is situated on the opposite side of the alimentary canal to the principal chain of ganglia of the nervous system; and communicates, by valvular apertures, with the surrounding venous sinus—the so-called "pericardium."

The *Crustacea* vary through such a wide range of organization that I doubt if any other anatomical proposition, in addition to those which I have mentioned, except the presence of a chitinous integument and the absence of cilia, can be enunciated, which shall be true of all the members of the group.

Fig. 27.

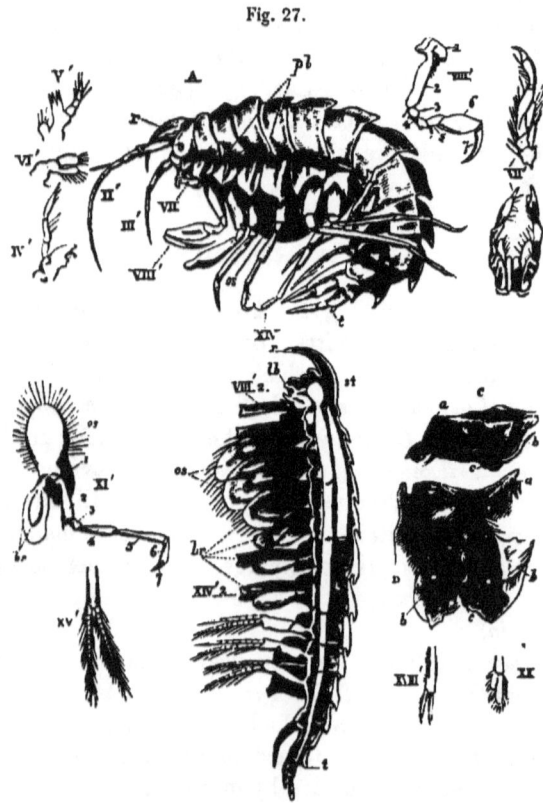

Fig. 27.—*Amphithoë*, an amphipodous Crustacean.—Lateral view (A), longitudinal and vertical section (B), detached appendages and stomach (C, D). The numbers I' to XX' indicate the appendages of the corresponding somites. *r*. Rostrum. *t*. Telson. *lb*. Labrum. *st*. Roof of the head, or cephalostegite. *os*. Oostegite. *Br*. Branchiæ. Stomach opened from above (D), and viewed laterally (C). *a, b, c*. Different parts of the armature.

XXI. THE ARACHNIDA.

It is this extreme elasticity, if I may so speak, of the crustacean type which renders the construction of any definition of the *Crustacea*, which shall include all its members and exclude the present class, so difficult. For the Spiders, Scorpions, Mites, and Ticks, which constitute this class, possess all the characters which have been just stated to be common to

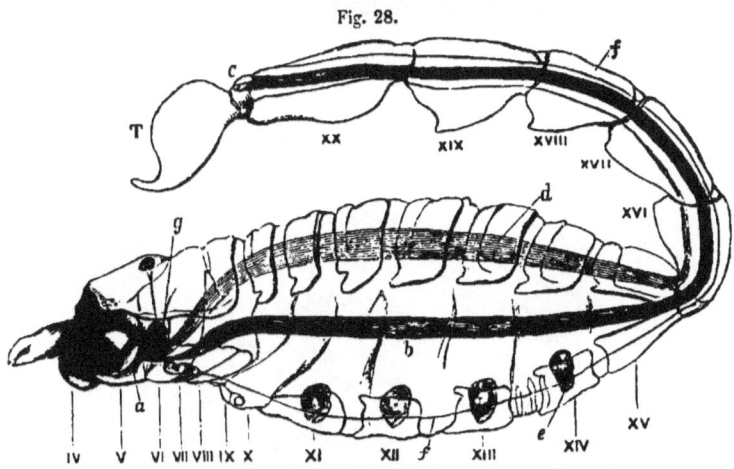

Fig. 28.—Diagrammatic section of a Scorpion, the locomotive members being cut away. *a*, Mouth leading into the pharyngeal pump. The large labrum lies above the mouth, and at the side of it are the bases of the large chelæ, or mandibles, IV., and above them the cheliceræ, or anteunæ. VI. to XX. Somites of the body. T, Telson; *b*, intestine; *c*, anus; *d*, indicates the position of the heart. *e*, the pulmonary sacs; *f*, a line indicating the position of the ganglionic chain; *g*, the cerebral ganglia.

the *Crustacea* save one; when they are provided with distinct respiratory organs, in fact, these are not external branchiæ, adapted for breathing aërated water, but are a sort of involution of the integument in the form of tracheal tubes, or pulmonary sacs, fitted for the breathing of air directly. But then many of the lower *Arachnida*, like the lower *Crustacea*, are devoid of special respiratory organs, and so the diagnostic character fails to be of service.

The following common characters of the *Arachnida*, however, help out our diagnosis in practice. They never possess more than four pairs of locomotive limbs, and the somites of the abdomen, even when the latter is well developed, are not provided with limbs. Again, in the higher *Arachnida* (Fig. 28), as in the higher *Crustacea*, the body is composed of twenty somites, six of which are allotted to the head; but, in the former class, one of the two normal pairs of antennæ is never developed, and the eyes are always sessile, while, in the highest *Crustacea*, the eyes are mounted upon moveable peduncles, and both pairs of antennæ are developed.

XXII. THE MYRIAPODA.

The Centipedes and Millipedes (Fig. 29) have the chitinous integument of the body divided into somites, provided with

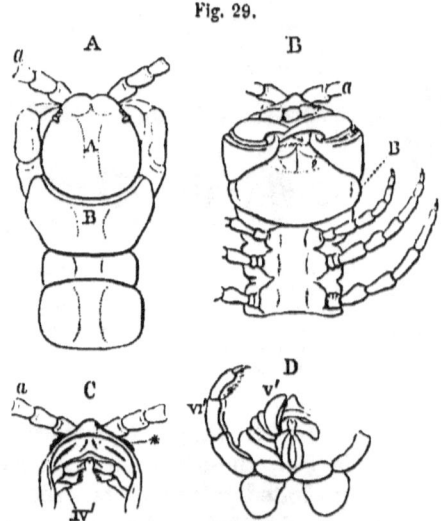

Fig. 29.—Anterior part of the body of *Scolopendra Hopei* (after Newport).—A, Anterior part of the body from above; B, from below; A, head proper; B, anterior thoracic somites; *a*, antennæ; C, antennæ, labrum, and mandibles (IV') from below; D, under view of head, with the two pairs of maxillæ (V' VI') covering the foregoing.

articulated appendages; and nervous and circulatory organs constructed upon a similar plan to those of the former groups.

THE MYRIAPODA.

The body consists of more than twenty somites, and those which correspond with the abdomen of *Arachnida* are provided with locomotive limbs.

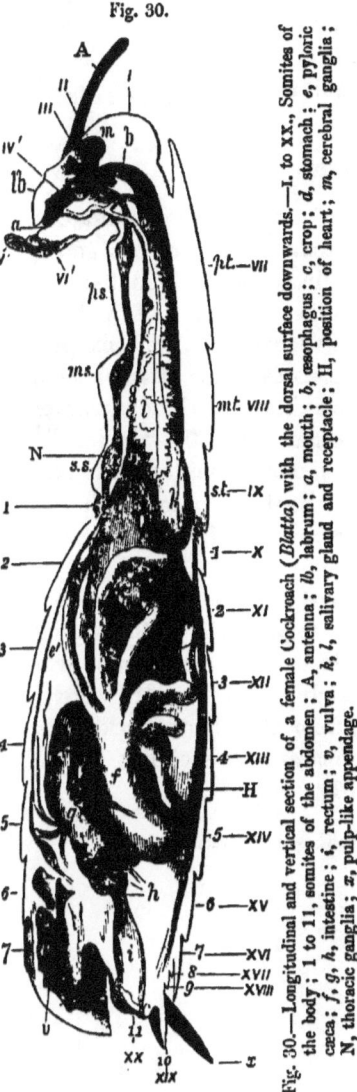

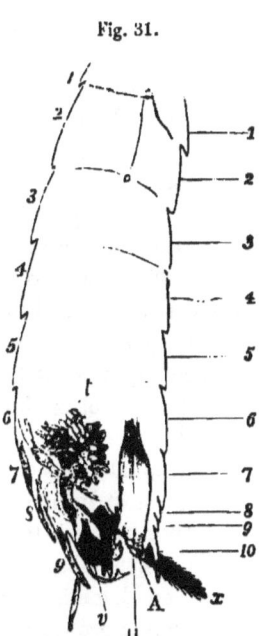

Fig. 30.—Longitudinal and vertical section of a female Cockroach (*Blatta*) with the dorsal surface downwards.—I. to XX., Somites of the body; 1 to 11, somites of the abdomen; A, antenna; *lb*, labrum; *a*, mouth; *b*, oesophagus; *c*, crop; *d*, stomach; *e*, pyloric caeca; *f, g, h*, intestine; *i*, rectum; *v*, vulva; *k, l*, salivary gland and receptacle; H, position of heart; *m*, cerebral ganglia; N, thoracic ganglia; *x*, pulp-like appendage.

Fig. 31.—Longitudinal and vertical section of the abdomen of a male Cockroach (*Blatta*).—1, 2, 3, 4, &c., terga and sterna of the abdomen; *t*, testis; *v*, aperture of the vas deferens; A, anus.

The head consists of at least five, and probably of six, coalescent and modified somites, and some of the anterior segments of the body are, in many genera, coalescent, and have their appendages specially modified to subserve prehension. The respiratory organs are tracheæ, which open by stigmata upon the surface of the body, and the walls of which are strengthened by chitin, so disposed as readily to pull out into a spirally coiled filament.

XXIII. The Insecta.

In this enormous assemblage of animals the respiratory organs are like those of the *Myriapoda*, with a nervous and a circulatory system disposed essentially as in this and the two preceding classes. But the total number of somites of the body never exceeds twenty. Of these five certainly, and six probably, constitute the head, which possesses a pair of antennæ, a pair of mandibles, and two pairs of maxillæ; the hinder pair of which are coalescent, and form the organ called the "labium."

Three, or perhaps, in some cases, more, somites unite and become specially modified to form the thorax, to which the three pairs of locomotive limbs, characteristic of perfect insects,* are attached.

Two additional pairs of locomotive organs—the wings—are developed, in most insects, from the tergal walls of the second and third thoracic somites. No locomotive limbs are ever developed from the abdomen of the adult insect, but the ventral portions of the abdominal somites, from the eighth backwards, are often metamorphosed into apparatuses ancillary to the generative function (Figs. 30 and 31).

* The female *Stylops* is stated to possess no thoracic limbs.

CHAPTER III.

THE CHARACTERS OF THE CLASSES OF THE VERTEBRATA.

THE five groups of animals which pass under the name of *Vertebrata*—the classes *Pisces, Amphibia, Reptilia, Aves,* and *Mammalia*—are obviously united and bound together by many common characteristics, and are well known to be so connected. In order to economise time and space, therefore, I shall preface my account of the character of these classes by enumerating the most important of those structural peculiarities which these five great divisions exhibit in common.

In the animals to which our attention has hitherto been confined, the external, or integumentary and parietal, portion of the blastoderm never becomes developed into more than a single saccular, or tubular, investment, which incloses all the viscera. So that if we make a transverse section of any one of these animals endowed with a sufficiently high organization to possess a nervous system and a heart, that section may be represented diagrammatically as in Fig. 32 (I.) where P represents the parietes or wall of the body, A the alimentary canal, H the heart, and N the nervous centres. It will be observed that the alimentary canal is in the middle, the principal centres of the nervous system upon one side of it, and the heart upon the other. In none of these animals, again, would you discover, in the embryonic state, any partition, formed by the original external parietes of the body, between the nervous centres and the alimentary canal.

But, in the five vertebrate classes, the parietal portion of the blastoderm of the embryo always becomes raised up, upon each side of the middle line, into a ridge, so that a long groove is

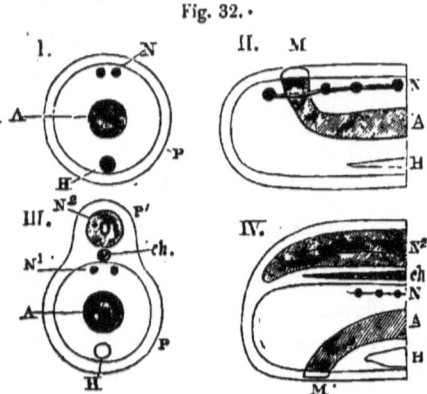

Fig. 32.—Diagrams representing generalised sections of one of the higher Invertebrates (I. II.), and of a Vertebrate (III. IV.); I. III. transverse, II. IV. longitudinal section. A, alimentary canal; H, heart; P, parietes of the body; P', parietes of the neural canal; N, nervous centres of Invertebrate; N^1, sympathetic, and N^2, cerebro-spinal centres of Vertebrate; ch, notochord; M, mouth.

formed between the parallel ridges thus developed; and the margins of these, eventually uniting with one another, constitute a second tube parallel with the first, by a modification of the inner walls of which the vertebrate cerebro-spinal nervous centres are developed. Hence it follows that, after any vertebrated animal has passed through the very earliest stages of its development, it is not a single, but a double tube, and the two tubes are separated by a partition which was, primitively, a part of the external parietes of the body, but which now lies, in a central position, between the cerebro-spinal nervous centres and the alimentary canal. Hence, a transverse section of any vertebrated animal may be represented diagrammatically by Fig. 32 (III.), where, for the most part, the letters have the same signification as in the foregoing case, but where P' denotes the second, or cerebro-spinal, tube. The visceral tube (P) contains, as in the case of the invertebrate animal, the alimentary canal, the heart, and certain nervous centres, belonging to the so-called sympathetic system. This nervous system and the heart are situated upon opposite sides of the alimentary canal, the sympathetic corresponding in position and in forming a double chain of ganglia, with the chief nervous centres of the invertebrate; so that the cerebro-spinal tube appears to be a super-addition,—a something not represented in the invertebrate series. The formation of the cerebro-spinal tube of vertebrates, in the manner which I have described to you, is a well-established fact; nor do I entertain any doubt that the cerebro-spinal centres, viz., the brain and the spinal cord of vertebrates, are the result of a modification of that serous layer of the germ which is continuous elsewhere with the epidermis. I have taken some pains to verify the remarkable discoveries of Remak in relation to this point, and, so far as the chick is concerned, his statements appeared to me to be fully borne out. But, as Von Baer long ago suggested, it is a necessary result of these facts that there can be no comparison between the cerebro-spinal nervous centres of the *Vertebrata* and the ganglionated nervous centres of the *Invertebrata*, and the homologues of the latter must probably be sought in the sympathetic.

Doubtless in close connection with this profound difference between the chief nervous centres of the vertebrate and the invertebrate is another remarkable structural contrast. In all the higher invertebrates, with a well-developed nervous system, the latter is perforated by the gullet, so that the mouth is situated upon the same side of the body as the principal masses of the nervous system, and some of the ganglia of the latter lie in front of, and others behind, the œsophagus. A longitudinal vertical section of such an animal, therefore, may be represented by Fig. 32 (II.).

A similar section of a vertebrated animal shows, on the contrary, the chief centre of the nervous system not to be perforated by the œsophagus; the latter turning away from it and opening upon the opposite side of the body (Fig. 32, IV.).

Another structure sharply distinctive of the vertebrate classes is the *chorda dorsalis* or *notochord*, an organ of which no trace has yet been discovered in any of the invertebrates, though it invariably exists, in early embryonic life at least, in every vertebrate. Before the cerebro-spinal canal is complete, in fact, the substance of the centre of its floor, beneath the primitive median line of the embryo, becomes differentiated into a rod-like cellular structure, which tapers to both its extremities; and, in a histological sense, remains comparatively stationary, while the adjacent embryonic tissues are undergoing the most rapid and varied metamorphoses.

To these great differences between vertebrates and invertebrates, in their early condition, many others might be added. In all *Vertebrata* that part of the wall of the body which lies at the sides of, and immediately behind the mouth, exhibits a series of thickenings parallel with one another and transverse to the axis of the body, which may be five, or more, in number, and are termed the "visceral arches." The interspaces between these arches becoming thinner and thinner, are at length perforated by corresponding clefts, which place the cavity of the pharynx in free communication with the exterior. Nothing corresponding with these arches and clefts is known in the *Invertebrata*.

A vertebrated animal may be devoid of articulated limbs, and it never possesses more than two pair. These limbs always have an internal skeleton, to which the muscles moving the limbs are attached. Whenever an invertebrated animal possesses articulated limbs, the skeleton to which the muscles are attached is external, or is connected with an external body skeleton.

When an invertebrated animal possesses organs of mastication, these are either hard productions of the alimentary mucous membrane, or are modified limbs. In the latter case there may be many pairs of them—numerous *Crustacea*, for example, have eight pairs of limbs devoted to this function. In no vertebrated animal, on the other hand, are limbs so modified and functionally applied, the jaws being always parts of the cephalic parietes specially metamorphosed, and totally distinct in their nature from the limbs. All vertebrated animals, finally, possess a distinct vascular system, containing blood with suspended corpuscles of one kind, or of two, or even three, distinct kinds. In all, save one, there is a single valvular heart—the vessels of the exception, *Amphioxus*, possessing numerous contractile dilatations. All vertebrates possess a "hepatic portal system," the blood of the alimentary canal never being wholly returned directly to the heart by the ordinary veins, but being more or less largely collected into a trunk, the "portal vein," which ramifies through and supplies the liver.

These are the most important characters by which the vertebrate classes are distinguished, as a whole, from the other classes of the animal kingdom ; and their number and importance go a long way to justify the step taken by Lamarck when he divided the animal kingdom into the two primary subdivisions of *Vertebrata* and *Invertebrata*.

XXIV. THE PISCES.

If we seek now to construct definitions of the first two classes of the *Vertebrata*, PISCES and AMPHIBIA, we shall meet with some difficulties, arising partly from the wide variations observable in the structure of fishes, and partly from the close affinity which exists between them and the *Amphibia*.

No fish exhibits any trace of that temporary appendage of the embryo of the higher vertebrates which is termed an amnion, nor can any fish be said to possess an allantois, though the urinary bladder of fishes may possibly be a rudiment of that structure. The posterior visceral clefts and arches* of fishes persist throughout life, and are usually more numerous than in other vertebrates; while upon, or in connection with, them are developed villi, or lamellæ, which subserve the respiratory function.

Median fins, formed by prolongations of the integument, supported by one or other kind of skeleton, are very characteristic of fishes ; and it is questionable if any fish exists altogether devoid of the system of median fin-rays and their supports, which have been termed inter-spinous bones and cartilages. On the other hand, no vertebrate animal, other than a fish, is known to possess them.

When the limbs, or pectoral and ventral fins, of fishes are

* The relation of the perforated pharynx of *Amphioxus* to the visceral arches and clefts is not known.

developed, they always exhibit a more or less complete fringe of fin-rays. No amphibian is known to possess such rays in its lateral appendages, but there is some reason to believe that the extinct *Ichthyosauria* may have been provided with them.

In most fishes, the nasal sacs do not communicate directly with the cavity of the mouth, but the *Myxinoids* and *Lepidosiren* are exceptions to this rule.

The blood-corpuscles of fishes are always nucleated, and are commonly red, but by a singular exception those of *Amphioxus* (the Lancelet, which is an exception to most rules of piscine organization) are colourless.

Almost all fishes have the heart divided into two chambers, one auricle and one ventricle; but *Amphioxus*, as I have previously stated, is devoid of any special heart, being provided instead with a number of contractile, vascular dilatations; while *Lepidosiren* possesses two auricles, and, at the same time, is provided with true lungs.

It is useless, therefore, to appeal to the olfactory organ, the blood, the heart, or the respiratory organs, for characters at once universally applicable to, and diagnostic of, fishes.

XXV. THE AMPHIBIA.

The Batrachians, Salamandroids, *Cæciliæ*, and Labyrinthodonts resemble fishes, and differ from all other vertebrates in the entire absence of an amnion, and in having only the urinary bladder to represent the allantois. They have red nucleated blood-corpuscles. Yet again, they resemble fishes and differ from all other vertebrates in the fact that filaments exercising a respiratory function, or branchiæ, are developed from their visceral arches during a longer or shorter period.

Some possess median fins, but these are not supported by fin-rays, and their limbs are never fringed with fin-rays.

Furthermore, in all *Amphibia* which possess limbs, the skeleton of these limbs is divisible into parts which obviously correspond with those found in the higher vertebrates. That is to say, in the fore limbs there are cartilages, or bones, answering in their essential characters and arrangement to the humerus, radius and ulna, carpus, metacarpus, and phalanges; and, in the hind limb, to the femur, tibia and fibula, tarsus, metatarsus, and phalanges of the higher vertebrates. This is the case in no fish; for, whether fishes possess parts corresponding with the humerus, radius and ulna, &c., or not, it is certain that the elements of their limb skeletons are very differently disposed from the arrangement which obtains in *Amphibia* and in higher vertebrates.

In all *Amphibia* the skull articulates with the spinal column by two condyles, and the basi-occipital remains unossified.

This is a character by which the *Amphibia* are sharply distinguished from the higher vertebrates.

There is a striking contrast between the close affinity of the fish and the amphibian and the wide separation of the *Amphibia* from the succeeding classes, all of which possess, in the embryonic state, a well-developed *amnion* and *allantois*, the latter almost always taking on, directly or indirectly, a respiratory function.

The amnion is a sac filled with fluid, which envelopes and shelters the embryo, during its slow assumption of the condition in which it is competent to breathe and receive food from without. The mode of its formation is shown in the accompanying figures of the early stages of development of the common fowl. Fig. 33, A, represents the first step in the differentiation of the embryo from the central portion of the blastoderm—that thin, membranous, cellular expansion which lies on the surface of the yelk where we see the cicatricula, or "tread." A well-defined, though shallow, straight groove, the "primitive groove," bounded at the sides by a slight elevation of the blastoderm, indicating the position of the future longitudinal axis of the body of the chick. Soon, the lateral boundaries of this groove, in what will become the anterior region of the body, grow up into plates—the dorsal laminæ (Fig. 33, B); and these dorsal laminæ, at length uniting, inclose the future cerebro-spinal cavity (Fig. 33, C, D). The blastoderm, beyond the region at which the dorsal laminæ are developed, grows downwards to form the ventral laminæ, and where the margins of these pass into the general blastoderm,

F

the outer, serous, or epidermic, layer rises up into a fold, which encircles the whole embryo; and the anterior and posterior parts

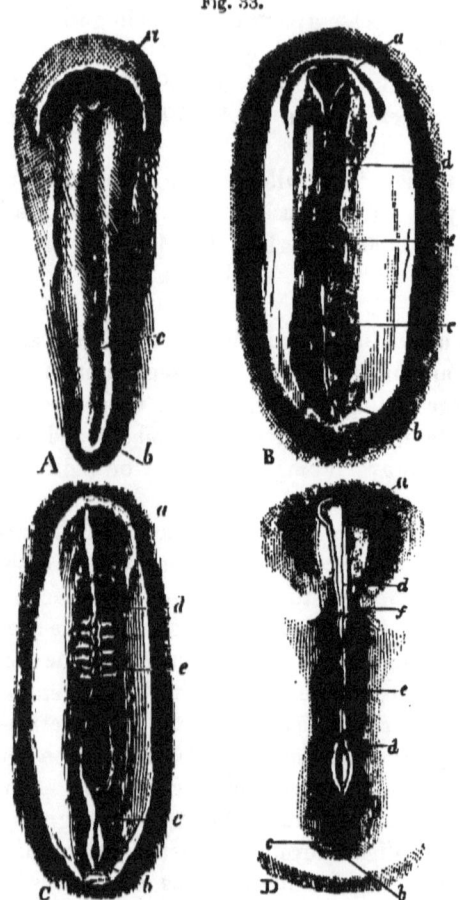

Fig. 33.—Development of the Chick.

A. First rudiment of the embryo; *a*, its cephalic; *b*, its caudal end; *c*, primitive groove.
B. The embryo further advanced; *a*, *b*, *c*, as before; *d*, the dorsal laminæ developed in the cephalic region only, and nearly uniting in the middle line; *e*, the proto-vertebræ.
C. Letters as before. The dorsal laminæ have united throughout the greater part of the cephalic region, and are beginning to unite in the anterior spinal region.
D. Embryo further advanced (second day), the dorsal laminæ having united throughout nearly their whole length. The proto-vertebræ have increased in number, and the omphalo-meseraic veins, *f*, are visible.

The embryos are drawn of the same absolute length, but it will be understood that the older embryos are, in nature, longer than the younger.

THE AMPHIBIA. 67

of this fold growing more rapidly than the lateral portions, form a kind of hood for the cephalic and caudal ends of the body respectively (Fig. 34, E). The margins of the hoods and of their lateral continuations at length meet over the middle line of the body, and there coalesce: so that the body is covered for a while by a double sac, the inner layer of which is formed by that wall of the fold of the serous layer which is inferior, or nearest to the body of the embryo; while the outer layer is formed by that wall which is superior, or furthest from the body of the embryo. The outer layer eventually disappears as a dis-

Fig. 34.

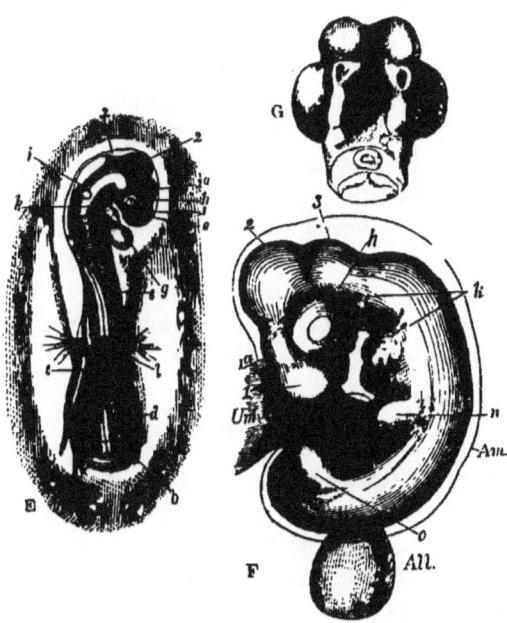

Fig. 34.—Development of the Chick.

E. Embryo at the third day; *g*, heart; *h*, eye; *i*, ear; *k*, visceral arches and clefts; *l, m*, anterior and posterior folds of the amnion, which have not yet united over the body; 1, 2, 3, first, second, and third cerebral vesicles; 1*a*, vesicle of the third ventricle.
F. Chick at the fifth day; *n, o*, rudiments of the anterior and posterior extremities; *Am*, amnion; *All*, allantois; *Um*, umbilical vesicle.
G. Under view of the head of the embryo F, the first visceral arch being cut away.

F 2

tinct structure, while the inner remains as the amnion. From the mode of formation which has been described, it results that the amnion is a shut sac, enveloping the body of the embryo; and is continuous, on the ventral side of the body, with the integument of a region which eventually becomes the umbilicus (Fig. 34, F).

The allantois is developed much later than the amnion, neither from the serous nor from the mucous (or epidermic and epithelial) layers of the germ, but from that intermediate stratum whence the bones, muscles, and vessels are evolved. It arises, as a solid mass, from the under part of the body of the embryo, behind the primitive intestinal cavity; and, enlarging, becomes a vesicle, which rapidly increases in size, envelopes the whole embryo, and, being abundantly supplied with arterial vessels from the aorta, serves as the great instrument of respiration during fœtal life; the porosity of the egg-shell allowing the allantoic blood to exchange its excess of carbonic acid for oxygen by osmosis.

The amnion and the external part of the allantois are thrown off at birth.

That which has just been stated respecting the development and characters of the amnion and allantois of the chick is true not only of all Birds, but of all *Reptilia*.

XXVI. THE REPTILIA.

All embryonic REPTILIA are provided with an amnion and an allantois, like those just described in the fœtal fowl. In the embryonic state, also, they possess visceral arches and clefts, but no respiratory tufts are ever developed in the arches, nor are reptiles endowed with an apparatus for breathing the air dissolved in water at any period of their existence. The skull of all *Reptilia* is articulated with the vertebral column by a single condyle, into which the ossified basi-occipital enters largely (Fig. 35). Each ramus of the lower jaw is composed of a number of pieces, and articulates with the skull, not directly, but by the intervention of a bone—the os quadratum—with which the hyoidean apparatus is not immediately connected (Fig. 36).

The fore-limb of Reptiles never takes the form of a wing, such as is seen in Birds; the "wing" of the remarkable extinct

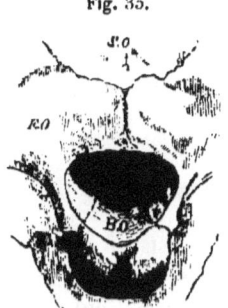

Fig. 35.—The occipital condyle of a Crocodile's skull viewed from behind.—*B.O*, Basioccipital; *E.O*, Ex-occipital; *S.O*, Supra-occipital.

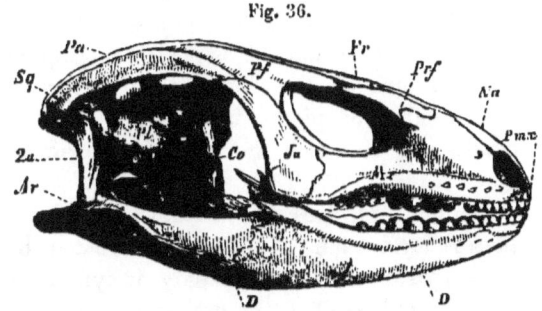

Fig. 36.—The skull of a Lizard (*Cyclodus*).—*D D*, Dentary piece of the lower jaw; *Ar*, Articular piece; *Qu*, Os quadratum; *Sq*, Squamosal.

flying reptiles, the *Pterodactyles*, being constructed on a totally different principle from that of a bird. In no known reptile are the metatarsal and tarsal bones ankylosed into one bone, except, perhaps, in the extinct genus *Compsognathus*.

In all *Reptilia* the greater and lesser circulations are directly connected together, within, or in the immediate neighbourhood of, the heart; so that the aorta, which is formed by the union of two arches, contains a mixture of venous and arterial blood.

The blood is cold, and the majority of the blood-corpuscles are red, oval, and nucleated. The bronchial tubes are not connected at the surface of the lungs with terminal saccular dilatations, or air-sacs. When, as is ordinarily the case, the superficial layers of the epidermis of Reptiles are converted into horn, the corneous matter takes the form of broad plates, or of overlapping scales, neither plates nor scales being developed within pouches of the integument.

XXVII. The Aves.

The class of Birds consists of animals so essentially similar to Reptiles in all the most essential features of their organization, that these animals may be said to be merely an extremely modified and aberrant Reptilian type.

As I have already stated, they possess an amnion and a respiratory allantois, and the visceral arches never develop branchial appendages. The skull is articulated with the vertebral column by a single condyle, into which the ossified basi-occipital enters largely. Each ramus of the lower jaw, composed, as in Reptiles, of a number of pieces, is connected with the skull by an os quadratum, to which the hyoidean apparatus is not suspended.

In no existing bird does the terminal division of the fore-limb possess more than two digits terminated by claws, and the metacarpal bones are commonly ankylosed together, so that the "manus" is of little use, save as a support for feathers.

In the hind limb of all birds the distal tarsal bone and the metatarsal bones become more or less completely ankylosed together, so as to form a single osseous mass, the "tarso-metatarsus."

The greater and lesser circulations of birds are completely separate, and there is only one aortic arch, the right. The right ventricle has a muscular valve. The blood is hot, hotter on the average than that of any other vertebrates, and the majority of the blood-corpuscles are oval, red, and nucleated. The bronchial tubes open upon the surface of the lungs into air-sacs, which differ in number and in development in different birds. Lastly, the integument of birds is always provided with horny appendages, which result from the conversion into horn of the cells of the outer layer of the epidermis. But the majority of these appendages, which are termed "feathers," do not take the form of mere plates developed upon the surface of the skin, but are evolved within sacs from the surfaces of conical papillæ of the dermis. The external surface of the dermal papilla, whence a feather is to be developed, is provided upon its dorsal surface with a median groove, which becomes shallower towards the apex of the papilla. From this median groove lateral furrows proceed at an open angle, and passing round upon the under surface of the papilla, become shallower, until, in the middle line, opposite the dorsal median groove, they become obsolete. Minor grooves run at right angles to the lateral furrows. Hence the surface of the papilla has the character of a kind of mould, and if it were repeatedly dipped in such a substance as a solution of gelatine, and withdrawn to cool until its whole surface was covered with an even coat of that substance, it is clear that the gelatinous coat would be thickest at the basal or anterior end of the median groove, at the median ends of the lateral furrows, and at those ends of the minor grooves which open into them; while it would be very thin at the apices of the median and lateral grooves, and between the ends of the minor grooves. If, therefore, the hollow cone of gelatine, removed from its mould, were stretched from within; or if its thinnest parts became weak by drying; it would tend to give way, along the inferior median line, opposite the rod-like cast of the median groove and between the ends of the casts of the lateral furrows, as well as between each of the minor grooves, and the hollow cone would expand into a flat feather-like structure with a median shaft, as a "vane" formed of "barbs" and "barbules." In point of fact, in the development of a feather such a cast of the dermal papilla is formed, though not in gelatine, but in the horny epidermic layer developed upon the mould, and, as this is thrust outwards, it opens out in the manner just described. After a certain period of growth the papilla of the feather ceases to be grooved, and a continuous horny cylinder is formed, which constitutes the "quill."

XXVIII. THE MAMMALIA.

All Mammals possess an amnion of an essentially similar character to that of Birds and Reptiles, and all have an allantois. But the latter either ceases to exist after a very early period of fœtal life, or else it is "placentiferous," and serves as the means of intercommunication between the parent and the offspring. Of the nature and characters of the "placenta" developed in the majority of the *Mammalia* I shall speak more particularly by and by. For the present, I pass it over as a structure not universally characteristic of the class.

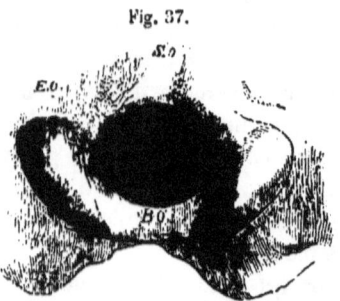

Fig. 37.

Fig. 37.—The occipital condyles of a Dog's skull viewed from behind.—Signification of the letters as in Fig. 35.

The visceral arches are, throughout life, as completely devoid of branchial appendages in Mammals, as in Birds and Reptiles. In the skull, the basi-occipital is well ossified; and, with the ex-occipitals, enters into the formation of the craniospinal articulation; the occipital condyle thus formed, however, is not single, as in Reptiles and Birds, but double, and the atlas has corresponding articular facets.

Each ramus of the lower jaw is composed of only a single piece, and this articulates directly with the squamosal bone of the skull, and not with the representative of the quadrate bone. The representative of the quadrate bone of the lower *Vertebrata* is appropriated, as *malleus*, to the service of the organs of hearing.

The brain possesses a *corpus callosum*.

The greater and lesser circulations of Mammals are as completely distinct as in Birds, and there is but a single aortic arch, the left. The majority of the blood-corpuscles are red, free nuclei, and these are always discoidal, and usually circular in form. The blood is hot. There is a complete diaphragm, and none of the bronchi end in air-sacs, like those of Birds.

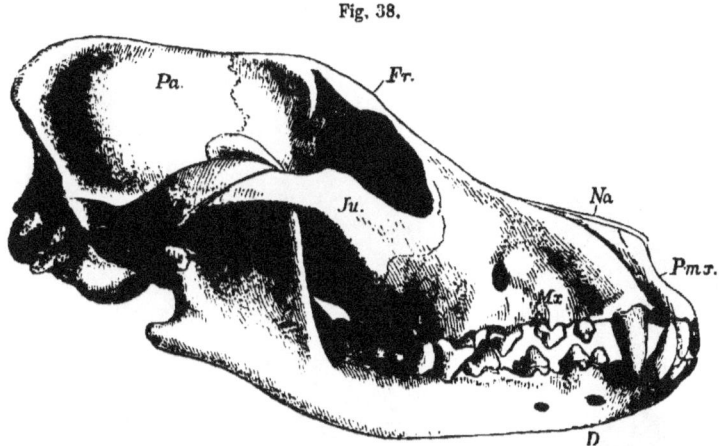

Fig. 38.—The skull of a Dog.—*D.* Ramus of the lower jaw; *Sq.* Squamosal.

Some part or other of the integument of all Mammals exhibits "hairs"—horny modifications of the epidermis—which so far resemble feathers, that they are developed upon papillæ inclosed within sacs; but, on the other hand, differ from the horny appendages of birds, in not splitting up as they are protruded, in the fashion so characteristic of feathers.

Finally, all Mammals are provided with organs for the secretion of a fluid which subserves the nourishment of the young after birth. The fluid is milk; the organs are the so-called "mammary" glands, and may probably be regarded as an extreme modification of the cutaneous sebaceous glands. These glands are aggregated into two or more masses, disposed upon each side of the median line of the ventral surface of the body; and, in almost all Mammals, the aggregated ducts of each mass open upon an elevation of the skin common to all—the nipple or teat. To this the mouth of the newly-born Mammal is applied, and from it, either by suction on the part of the young, or by the compressive action of a special muscle on the part of the parent, the nutritive fluid makes its way into the stomach of the former.

CHAPTER IV.

ON THE ARRANGEMENT OF THE CLASSES INTO LARGER GROUPS.

HAVING now arrived at the end of the list of classes, and obtained a comprehension of the structural features common to, and characteristic of, each class, it will be proper next to discuss the relations of these classes one to another, and to inquire how far they present such common characters as will enable us to group them into larger divisions.

And, to commence with the highest classes, it is clear that the *Mammalia*, *Aves*, and *Reptilia* are united together by certain very striking features of their development. All possess an amnion and an allantois, and are devoid, throughout life, of any apparatus for breathing the air which is dissolved in water. Thus they have been termed ABRANCHIATE VERTEBRATA, in contradistinction to *Pisces* and *Amphibia*, which possess no amnion, nor allantois (or at most a rudimentary one), and, being always provided at a certain period, if not throughout life, with branchiæ, have been called BRANCHIATE VERTEBRATA.

The Abranchiate, however, form a far less homogenous assemblage than the branchiate *Vertebrata*—Mammals being so strongly separated from Reptiles and Birds that I am disposed to regard them as constituting one of three primary divisions, or provinces, of the *Vertebrata*. The structure of the occipital condyles, the structure and mode of articulation of the mandibular rami, the presence of mammary glands, and the non-nucleated red blood-corpuscles appear to separate Mammals as widely from Birds and Reptiles as the latter are separated from *Amphibia* and Fishes.

Thus the classes of the *Vertebrata* are capable of being grouped into three provinces: (I.) the ICHTHYOPSIDA (comprising *Pisces* and *Amphibia*), defined by the presence of branchiæ at some period of existence, the absence of an amnion, the absence, or rudimentary development, of the allantois, and blood-corpuscles which are always nucleated; (II.) the SAUROPSIDA, comprising *Reptilia* and *Aves*, defined by the absence of branchiæ at all periods of existence, the presence of a well-developed amnion and allantois, a single occipital condyle, a complex mandibular ramus articulated to the skull by a quadrate bone, and nucleated red blood-corpuscles; and (III.) the MAMMALIA, devoid of branchiæ and provided with an amnion and an allantois, but with two occipital condyles and a well-developed basi-occipital; with a simple mandibular ramus articulated with the squamosal and not with the quadratum; with mammary glands; with red non-nucleated blood-corpuscles; and with a corpus callosum in the brain.

These five classes, whether divided into two or three provinces, again, present so many characters, already enumerated, by which they resemble one another, and differ from all other animals, that, by universal consent, they are admitted to form the group of VERTEBRATA, which takes its place as one of the primary divisions or "sub-kingdoms" of the Animal Kingdom.

The next four classes—*Insecta, Myriapoda, Arachnida, Crustacea*—without doubt also present so many characters in common as to form a very natural assemblage. All are provided with articulated limbs attached to a segumented body-skeleton—the latter, like the skeleton of the limbs, being an "exoskeleton," or a hardening of that layer which corresponds with the outer part of the epidermis of Vertebrates. In all, at any rate in the embryonic condition, the nervous system is composed of a double chain of ganglia, united by longitudinal commissures, and the gullet passes between two of these commissures. No one of the members of these four classes is known to possess vibratile cilia. The great majority of these animals have a distinct heart, provided with valvular apertures which are in communication with a perivisceral cavity containing corpusculated blood. But the *Cirripedia* and the *Ostracoda* among Crustaceans, and many of the Mites among *Arachnida*, have as yet yielded no trace of distinct circulatory organs, so that the nature of these organs cannot be taken as a universal character of the larger group we are seeking; still less can such a character be found in the respiratory organs, which vary widely in character, and are often totally absent as distinct structures. A striking uniformity of composition is to be found in the heads of, at any rate, the more highly organized members of these four classes, so that, typically, the head of a Crustacean, an Arachnid, a Myriapod, or an Insect is composed of six somites (or segments corresponding with those of the body) and their appendages, the latter being modified so as to serve the purpose of sensory and manducatory organs. I believe this doctrine to be substantially correct; and that, leaving all hypothetical suppositions aside, the head of any animal belonging to these classes may be demonstrated to contain never fewer than four, and never more than six somites with their appendages; but, until this view has received confirmation from other workers, I shall not venture to put forward any statement based upon it as part of the definition of the large group or "province" containing the four classes above mentioned, which has received from some naturalists the name of ARTICULATA, from others that of ARTHROPODA, the latter being perhaps the more distinctive and better appellation.

The members of the class *Annelida* present marked differences from all the *Arthropoda*, but resemble them in at least one important particular; and that is, the arrangement of the nervous system, which constitutes a ganglionated double chain, traversed at one point by the œsophagus. In almost all other respects, Annelids differ widely from Arthropods. It may be doubted whether any Annelid is devoid of cilia in some part or other of its organization, and cilia constitute the most important organs of locomotion in the embryos of many. No Annelid possesses a heart communicating by valvular apertures with the perivisceral cavity, none have articulated limbs, and none possess a head composed of even four modified somites.

Most Annelids are provided with that peculiar system of vessels termed "pseudo-hæmal;" but, in some, that system has not yet been discovered.

In endeavouring to separate from among invertebrated animals a first large group, comparable to the *Vertebrata*, it appears to me that the resemblances between the *Annelida*, the *Chætognatha*, and the *Arthropoda* outweigh the differences; and that the characters of the nervous system and the frequently segmented body, with imperfect lateral appendages, of the former, necessitate their assemblage with the *Arthropoda* into one great division, or "sub-kingdom," of ANNULOSA.

But what of the *Echinodermata* and the *Scolecida* ? Should both these great classes be also ranged under the *Annulosa*; or do they belong to different sub-kingdoms; or, if they belong to the same, should they constitute a sub-kingdom of their own?

I will endeavour to reply to these questions in succession. Whether these two groups belong to the *Annulosa* or not, must depend upon whether they possess any characters in common with the *Arthropoda* and *Annelida* other than those which they have in common with all animals. I can find none of any great moment. No Echinoderm, or Scolecid, has a definitely segmented body or bilaterally disposed successive pairs of appendages. None of these animals has a longitudinal chain of ganglia.

On the other hand, there is much resemblance between the ciliated larvæ of some Scolecids and Echinoderms, and those of Annelids; and the form of the body of many Scolecids is so similar to that of one of the most familiar of Annelids, as to have earned for both them and the Annelids the common title of "worms." Nor must it be forgotten that, in the Annelids, there seem to be representatives of that singular system of vessels which attains so large a development as the "water-vascular" apparatus in many Scolecids.

Whatever value may be attached to these resemblances, it must, I think, be admitted that, in the present state of our knowledge, it is impossible to affirm anything absolutely common to, and yet diagnostic of, all *Annulosa* and all Echinoderms and Scolecids. On the other hand, there can be no doubt as to the many and singular resemblances which unite the Scolecids and the Echinoderms together. The nervous system of the Echinoderm may present considerable differences from that of a Trematode or Rotifer, but it must be recollected that the comparison is not a fair one, seeing that the mouth and gullet of an Echinoderm, round which its nervous ganglia are arranged, are not, strictly speaking, the same as the parts so named in a Rotifer, but are new developments.

And it is exactly in that anomalous method of development of the Echinoderm within its larva, which is so characteristic of the whole group of *Echinodermata*, that this class exhibits its strong alliance with the *Scolecida*; the *Turbellaria* and *Tæniada* exhibiting the only approach to the method of Echinoderm development known in the Animal Kingdom.

A singular larva studied by Johannes Müller, in one of his many fruitful visits to the seashore, and termed by him *Pilidium*, has furnished, in the hands of subsequent observers (more especially Krohn, Leuckart, and Pagenstecher), ample proof that a *Nemertes* (a genus of *Turbellaria*) may be developed in a manner altogether similar to that in which an Echinoderm takes its origin.

The *Pilidium* (Fig. 39) is a small, helmet-shaped larva, with a long flagellum attached like a plume to the summit of the helmet, the edges and side lobes of which are richly ciliated. A simple alimentary sac opens upon the under surface of the body between the lobes (Fig. 39, A).

In this condition, the larva swims about freely; but, after a while, a mass of formative matter appears upon one side of the alimentary canal, and, elongating gradually, takes on a wormlike figure. Eventually it grows round the alimentary canal, and, appropriating it, detaches itself from the *Pilidium* as a Nemertid—provided with the characteristic proboscis, and the other organs of that group of *Turbellaria*.

Many *Trematoda*, and all Tænioid *Scolecida*, again, present an essentially similar process of internal gemmation, in virtue of which either a separate offspring arises, or an adult is developed within an embryonic form; but in these cases the

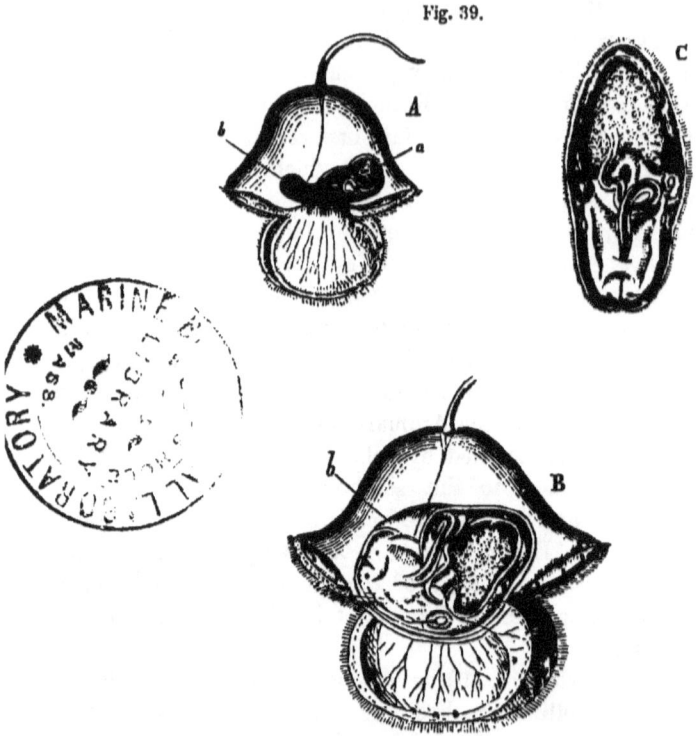

Fig. 39.—*Pilidium gyrans* (after Leuckart and Pagenstecher).
A. Young *Pilidium*: *a*, alimentary canal; *b*, rudiment of the Nemertid.
B. *Pilidium* with a more advanced Nemertid.
C. Newly-freed Nemertid.

appropriation of the intestine of the primary by that of the secondary form, which renders the ordinary development of the Echinoderm so striking, does not occur.

In discussing the characters of the *Echinodermata*, I have described at length the ambulacral system; and, in speaking of the *Scolecida*, I have no less insisted upon the peculiarities of the "water-vascular system." But it is impossible to compare these two systems of vessels without being struck by their similarity. Each is a system of canals, opening externally, and ciliated within; and the circumstance that the two apparatuses are turned to different purposes in two distinct groups of the animal kingdom, seems to me no more to militate against their homology, than the respiratory function of the limbs of the Phyllopod *Crustacea* militates against the homology of these limbs with the purely locomotive appendages of other Crustaceans.

Thus it appears that the *Echinodermata* and the *Scolecida* are so closely connected that they can by no means be placed in separate sub-kingdoms; and, in the course of studying the other sub-kingdoms, it will be quite obvious that, unless they are to occupy an independent position, there is no place for them anywhere, save among the *Annulosa*. I have hitherto been accustomed to consider them, under the name of the ANNULOIDA, as a division of this sub-kingdom; but until some structural character can be discovered by which all the *Annuloida* agree with the *Annulosa*, and differ from other animals, I am much inclined to think it would conduce to the formation of clear conceptions in zoology if the *Annuloida* were regarded as a distinct primary division of the Animal Kingdom.

If we now turn to the other column of classes of invertebrate animals (p. 6), the last four on the list, viz., *Cephalopoda, Pteropoda, Pulmogasteropoda,* and *Branchiogasteropoda*, have a number of well-marked characters in common. In all, the nervous system is composed of three principal pairs of ganglia —cerebral, pedal, and parieto-splanchnic—united by commissures. All possess that remarkable buccal apparatus, the odontophore,—whence I have ventured to propose the name of ODONTOPHORA for the group. The circulatory and respiratory organs vary a great deal, but none are provided with double lamellar gills upon each side of the body.

The *Lamellibranchiata* stand in somewhat the same relation to the *Odontophora* as the *Annelida* to the *Arthropoda*. The Lamellibranchs have the three fundamental pairs of ganglia of the *Odontophora*, but they possess no trace of the odontophore. Furthermore, they are all provided with bivalve external pallial shells, the valves being right and left in relation to the body.

No shell of this kind is found in any of the *Odontophora*. Almost all Lamellibranchs, lastly, have a pair of lamellar gills on each side of the body, and all are provided with auriculate hearts. No doubt the *Odontophora* and the *Lamellibranchiata* properly form parts of one and the same sub-kingdom, MOLLUSCA, and the three classes which follow, viz., the *Ascidioida*, *Brachiopoda*, and *Polyzoa*, are usually included in the same subkingdom.

But the difficulty of framing a definition which shall include the last-named classes with the *Lamellibranchiata* and *Odontophora* is almost as great as in the parallel case of the *Annuloida* and *Annulosa*; while, on the other hand, the Ascidians, Brachiopods, and *Polyzoa* exhibit many features in common. Thus the nervous system is greatly simplified in all three classes, consisting, in the *Ascidioida* and *Polyzoa*, of a single ganglion, sending perhaps a commissural cord round the gullet. In the *Brachiopoda* the chief ganglia, which appear to be the homologues of the pedal ganglia of the higher mollusks, and are connected by a circumœsophageal cord, are combined with accessory ganglia, but these do not seem to be identifiable with the pedal or the parieto-splanchnic ganglia.

Again, the fact that the heart, when present, is of a simple tubular, or saccular, character, and is devoid of any separation into auricle and ventricle, constitutes a wide difference between these three classes and the higher Mollusks. On the other hand, these classes, which may be conveniently denominated MOLLUSCOIDA, resemble one another in the fact (so far as I am aware there is only one exception, *Appendicularia*) that the mouth is provided with ciliated tentacula, disposed in a circle, or in a horse-shoe shape, or fringing long arms; that it leads into a large, and sometimes an exceedingly large, pharynx; and that in two of the three, at least, that system of cavities communicating with the exterior, which has been called the "atrial system," is greatly developed.

I cannot doubt, then, that the *Molluscoida* form a natural assemblage; but, until the precise characters, if any exist, which unite them with the *Mollusca* proper can be clearly defined, I am inclined to think it might be better, as in the case of the

Annuloida, to recognise them as a separate division of the Animal Kingdom.

The next two classes—the *Actinozoa* and the *Hydrozoa*—constitute one of the most natural divisions of the Animal Kingdom—the CŒLENTERATA of Frey and Leuckart. In all these animals, the substance of the body is differentiated into those histological elements which have been termed cells, and the latter are primarily disposed in two layers, an external and an internal, constituting the "ectoderm" and "endoderm."

Among animals which possess this histological structure, the *Cœlenterata* stand alone, in having an alimentary canal, which is open at its inner end and communicates freely, by means of this aperture, with the general cavity of the body. In a large proportion of these animals the prehensile organs are hollow tentacles, disposed in a circle around the mouth, and all (unless the *Ctenophora* should prove to be a partial exception to the rule) are provided with very remarkable organs of offence and defence, termed "thread cells" or "nematocysts." These, when well exhibited, as, for example, by the common freshwater polype (*Hydra*), are oval, elastic sacs, containing a long coiled filament, barbed at its base, and serrated along the edges. When fully developed, the sacs are tensely filled with fluid, and the slightest touch is sufficient to cause the retroversion of the filament, which then projects beyond the sac for a distance, which is not uncommonly equal to many times the length of the latter. These fine filaments readily penetrate any delicate animal tissue with which they are brought into contact, and cause great irritation in the human skin when they are of large size. Nor can it be doubted that they exert a similarly noxious influence upon the aquatic animals which are seized by, and serve as prey to, the *Actinozoa* and *Hydrozoa*. Characteristic as these organs are of the Cœlenterates, however, it must not be imagined that they are absolutely peculiar to the sub-kingdom; for some nudibranchiate *Mollusca*, such as *Eolis*, are armed with similar weapons, and the integument of certain *Turbellaria*, and even of some *Infusoria*, is provided with bodies which seem to be of a not altogether dissimilar character.

No Cœlenterate possesses any circulatory organs, unless the

cilia which line the general cavity of the body can be regarded as such; and a nervous system has, at present, been clearly made out only in the *Ctenophora*. Here its central mass occupies a position which is very unlike that in which the principal masses of the central nervous system are found in other invertebrate animals, being situated upon that side of the body which is diametrically opposed to the mouth.

Whatever extension our knowledge of the nervous apparatus of the Cœlenterates may, and not improbably will, receive from future investigators, the positive characters afforded by the histological features of their substance, and the free opening of their alimentary canal into the general cavity of the body, are such as to separate them, as a sub-kingdom, as sharply defined and devoid of transitional forms as that of the *Vertebrata*, from the rest of the Animal Kingdom.

Great difficulties stand in the way of any satisfactory grouping of the remaining classes, if we are determined to remain true to the principle that the definition of a group shall hold good of all members of that group, and not of any others,— a principle which lies at the foundation of all sound classification.

In possessing cilia, as locomotive and ingestive organs; in being provided with a contractile water receptacle with canals proceeding from it (in some cases at any rate) into the substance of the body; in their tendency to become encysted and assume a resting condition, the INFUSORIA undoubtedly exhibit analogies with the lower *Annuloida*, such as the *Turbellaria*, *Rotifera*, and *Trematoda*.

But the entire absence, so far as our present knowledge goes, of a nervous system, the abrupt termination of the gullet in a central semi-fluid sarcodic mass, and the very peculiar characters of the reproductive organs, of the *Infusoria*, separate them widely from the *Annuloida*, though it seems to me not improbable that the gap may hereafter be considerably diminished by observation of the lower forms of *Turbellaria*.

At present the *Infusoria* are usually regarded as forming part of the same sub-kingdom as the *Spongida*, *Radiolaria*, *Rhizopoda*, and *Gregarinida*, and as closely allied to them. But, so far as I am aware no definition can be framed which will yield characters at once common to, and distinctive of, all these four groups; while recent discoveries tend to widen so greatly the hiatus between the *Infusoria* and the other three classes, that I greatly doubt if the sub-kingdom PROTOZOA can be retained in its old sense.

But if the *Infusoria* be excluded from it, the remaining groups, notwithstanding the imperfection of our knowledge regarding some of them, exhibit a considerable community of partly negative and partly positive characters.

The *Spongida, Radiolaria, Rhizopoda,* and *Gregarinida*, in fact, are all devoid of any definite oral aperture; a considerable extent, and sometimes the whole, of the outer surface of the body acting as an ingestive apparatus. Furthermore, the bodies of these animals, or the constituent particles of the compound aggregations, such as the Sponges, exhibit incessant changes of form—the body wall being pushed out at one point and drawn in at another—to such an extent, in some cases, as to give rise to long lobate, or filamentous, processes, which are termed "pseudopodia."

Finally, all these classes agree in the absence of any well-defined organs of reproduction, innervation, or blood circulation.

Thus, in the present state of knowledge, it seems to me that the whole Animal Kingdom cannot be divided into fewer than eight primary groups, no two of which are susceptible, in the present state of knowledge, of being defined by characters which shall be at once common and diagnostic.

These groups are the—

VERTEBRATA.

MOLLUSCA.   ANNULOSA.
MOLLUSCOIDA.   ANNULOIDA.
CŒLENTERATA.   INFUSORIA.

PROTOZOA.

I leave aside altogether the question of the equivalency of these groups; and, as I have already stated, I entertain some doubts regarding the permanency of one—the *Infusoria*—as a distinct primary division. Nor, in view of the many analogies between the *Mollusca* and the *Molluscoida*, the *Annulosa* and the *Annuloida*, do I think it very improbable that, hereafter, some common and distinctive characters may possibly be discovered which shall unite these pairs respectively. But the discoveries which shall effect this simplification have not yet been made, and our classification should express not anticipations, but facts.

I have not thought it necessary or expedient, thus far, to enter into any criticism of the views of other naturalists, or to point out in what respect I have departed from my own earlier opinions. But Cuvier's system of classification has taken such deep root, and is so widely used, that I feel bound, in conclusion, to point out how far the present attempt to express in a condensed form the general results of comparative anatomy departs from that embodied in the opening pages of the "Règne Animal."

The departure is very nearly in the ratio of the progress of knowledge since Cuvier's time. The limits of the highest group, and of the more highly organized classes of the lower divisions, with which he was so well acquainted, remain as he left them; while the lower groups, of which he knew least, and which he threw into one great heterogeneous assemblage,—the *Radiata*,—have been altogether remodelled and rearranged. Milne-Edwards demonstrated the necessity of removing the *Polyzoa* from the radiate mob, and of associating them with the lower Mollusks. Frey and Leuckart demonstrated the subregnal distinctness of the *Cœlenterata*. Von Siebold and his school separated the *Protozoa*, and others have completed the work of disintegration by erecting the *Scolecida* into a primary division, of *Vermes*, and making the *Echinodermata* into another. Whatever form the classification of the Animal Kingdom may eventually take, the Cuvierian *Radiata* is, in my judgment, effectually abolished: but the term is still so frequently used, that I have marked out those classes which it denoted in the diagram of the Animal Kingdom (p. 6), so that students may not be at a loss to understand the sense in which it is employed.

CHAPTER V.

THE SUBCLASSES AND ORDERS INTO WHICH THE CLASSES OF THE VERTEBRATA ARE DIVISIBLE.

I. THE MAMMALIA.

A GREAT many systems of classification of the *Mammalia* have been proposed, but, as any one may imagine from the nature of the case, only those which have been published within the last forty or fifty years, or since our knowledge of the anatomy of these animals has approached completeness, have now any scientific standing-ground. I do not propose to go into the history of those older systems, which laboured more or less under the disqualification of being based upon imperfect knowledge, but I shall proceed, at once, to that important step towards dividing the *Mammalia* into large groups, which was taken by the eminent French anatomist, M. de Blainville, so far back as the year 1816. M. de Blainville pointed out that the *Mammalia* might be divided into three primary groups, according to the character of their reproductive organs, especially the reproductive organs of the female. He divided them into "Ornithodelphes," "Didelphes," "Monodelphes;" or, as we might term them, ORNITHODELPHIA, DIDELPHIA, MONODELPHIA. Now, I do not mean to assert that M. de Blainville defined these different groups in a manner altogether satisfactory, or strictly in accordance with all the subsequently discovered facts of science, but his great knowledge and acute intuition led him to perceive that the groups thus named were truly natural divisions of the *Mammalia*. And the enlargement of our knowledge by subsequent investigation seems to me, in the main, only to have confirmed De Blainville's views.

The subclass of the ORNITHODELPHIA comprises those two remarkable genera of Mammals, as isolated in geographical distribution as in structure,—*Ornithorhynchus* and *Echidna*—which constitute the order *Monotremata*.

In these animals the angle of the lower jaw is not inflected, and the jaws are devoid of true teeth, one of the two genera only (*Ornithorhynchus*) possessing horny plates in the place of teeth. The coracoid bone extends from the scapula to the sternum, with which it is articulated, as in birds and most reptiles, and, as in many of the latter, there is an inter-clavicle. There is no marsupial pouch, though bones wrongly termed "marsupial" are connected with the pelvis. But it is to the structure of the female reproductive organs that the *Ornithodelphia* owe their name. The oviducts, enlarged below into uterine pouches, but opening separately from one another, as in oviparous vertebrates, debouch, not into a distinct vagina, but into a cloacal chamber, common to the urinary and genital products and to the fæces. The testes of the male are abdominal in position throughout life, and the vasa deferentia open into the cloaca, and not into a distinct urethral passage. The penis is indeed traversed by a canal, but it is open and interrupted at the root of that organ. In both sexes, the ureters pour the renal secretion, not into the bladder, which is connected with the upper extremity of the cloaca, but into the latter cavity itself.

In the brain, the *corpus callosum* is small, the anterior commissure large. We are but very imperfectly acquainted with the reproductive processes of these animals, but it is asserted that the young are devoid of a placenta. The mammary gland has no nipple.

Like the *Ornithodelphia*, the subclass DIDELPHIA contains but a single order, the *Marsupialia*, the great majority of which, like the *Ornithodelphia*, inhabit Australia. They almost all have the angle of the lower jaw inflected, and all possess true teeth. The coracoid is, as in the higher Mammals, ankylosed with the scapula, and is not articulated with the sternum. All have the so-called "marsupial" bones or cartilages—ossifications, or chondrifications, of the internal tendon of the external oblique muscle of the abdomen—and the females of almost all possess a fold of the skin of the abdomen above the pubis, constituting a "*marsupium*," or pouch, within which the young are nourished and protected in their early, helpless condition.

The oviducts open into vaginæ, which are more or less completely divided into two separate passages. The testes of the males are lodged in a scrotum, which is suspended in front of the penis; and the vasa deferentia open into a complete and continuous urethra, which is also the passage by which the urine escapes from the bladder, and is perfectly distinct from the passage for the fæces, though the anus and the termination of the urethro-sexual canal are embraced by the same sphincter.

The corpus callosum is comparatively small, and the anterior commissure large, as in the *Ornithodelphia*.

It is stated that the allantois of the embryo is arrested in its development, and gives rise to no placenta. The umbilical sac is said to acquire a large proportional size; but whether it plays the part of a placenta for the short period of intra-uterine life, or not, is unknown.

The young are born of very small size, and in a singularly imperfect condition; but being transferred to the marsupium, and becoming attached to a long nipple, they are supplied with milk until they are able to provide for themselves—the milk being, at first, forced into their mouths by the action of a muscle spread over the mammary gland.

In the MONODELPHIA, the angle of the lower jaw is not inflected, and they may or may not be provided with teeth. They never possess "marsupial" bones. The uterine dilatation of the oviducts is always considerable, and whether they have common or distinct apertures, the vagina is a single tube, though it may be partially divided by a septum. The testes may vary much in position; but, if they are lodged in a scrotal pouch, it is never pendulous by a narrow neck in front of the penis, as in the *Didelphia*.

The urinary bladder opens into a distinct urethra, which, directly or indirectly, receives the vasa deferentia in the male.

The corpus callosum is very variable in its development, commonly attaining a much larger size than in the preceding groups; the optic lobes are divided into four portions.

The young are nourished within the uterus until such time as they are competent to suck milk from the teats of the parent, to which end the chorion always develops processes or villi, which are well supplied with vessels brought to them by the allantois. These processes becoming interlaced more or less closely with corresponding vascular developments of the wall of the uterus (and so forming a "placenta"), an interchange of constituents takes place between the fœtal and the maternal blood, through the separating walls of the fœtal and maternal vessels. In this manner, throughout its prolonged intra-uterine life, the Monodelphian fœtus is supplied with nourishment and gets rid of its effete products.

It is a well-established fact that two very distinct types of

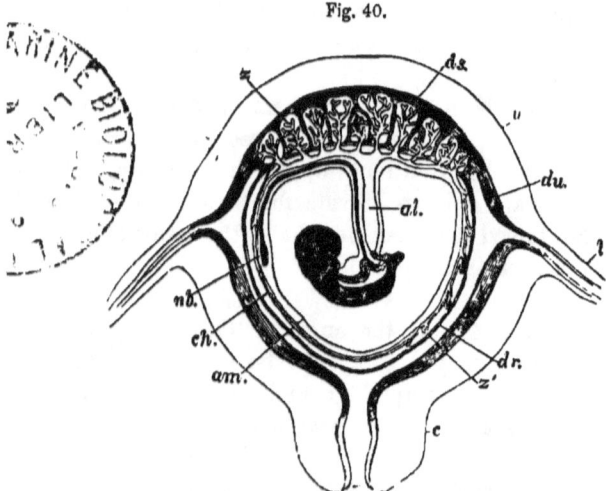

Fig. 40.

Fig. 40.—Diagrammatic section of a human pregnant uterus, with the contained ovum (Longet). *u*, uterus; *l*, oviduct; *c*, cervix uteri; *du*, decidua uteri; *dr*, decidua reflexa; *ds*, decidua serotina; *ch*, chorion; *am*, amnion; *al*, allantois; *nb*, umbilical vesicle; *z*, villi which form the fœtal part of the placenta; *z'*, villi over the rest of the chorion, which take no part in the placental function in man.

placenta are to be met with in the *Monodelphia*, and that, at the present moment, we have no knowledge of any transitional forms between these two types. The first of these types is that exhibited by the human placenta, the second by that of the pig or horse.

From the commencement of gestation, the superficial substance of the mucous membrane of the human uterus undergoes a rapid growth and textural modification, becoming converted into the so-called "*decidua.*" While the ovum is yet small, this *decidua* is separable into three portions,—the *decidua vera*, which lines the general cavity of the uterus; the *decidua reflexa*, which immediately invests the ovum; and the *decidua serotina*, a layer of especial thickness, developed in contiguity with those chorionic villi which persist and become converted into the fœtal placenta. The *decidua reflexa* may be regarded as an outgrowth of the *decidua vera*; the *decidua serotina* as a special development of a part of the *decidua vera*. At first, the villi of the chorion are loosely implanted into corresponding depressions of the *decidua*; but, eventually, the chorionic part of the placenta becomes closely united with, and bound to, the uterine decidua, so that the fœtal and maternal structures form one inseparable mass.

In the meanwhile, the deeper substance of the uterine mucous membrane, in the region of the placenta, is traversed by numerous arterial and venous trunks, which carry the blood to and from the placenta; and the layer of decidua into which the chorionic villi do not penetrate acquires a cavernous, or cellular, structure from becoming burrowed, as it were, by the innumerable sinuses into which these arterial and venous trunks open. In the process of parturition, the *decidua serotina* splits through this cellular layer, and the superficial part of it comes away with the unbilical cord, together with the fœtal membranes and the rest of the *decidua*; while the deeper layer, undergoing fatty degeneration and resolution, is more or less completely brought away with the *lochia*, and gives place to a new mucous membrane, which is developed throughout the rest of the uterus, during pregnancy; but, possibly, arises only after delivery over the placental area.

92 INTRODUCTION TO CLASSIFICATION.

In the Pig the placenta is an infinitely simpler structure. No "*decidua*" is developed; the elevations and depressions of the unimpregnated uterus simply acquire a greater size and vascularity during pregnancy, and cohere closely with the chorionic villi, which do not become restricted to one spot, but are developed from all parts of the chorion, except its poles, and remain persistent in the broad zone thus formed throughout foetal life. The cohesion of the foetal and maternal placentæ, however, is overcome by slight maceration or post-mortem

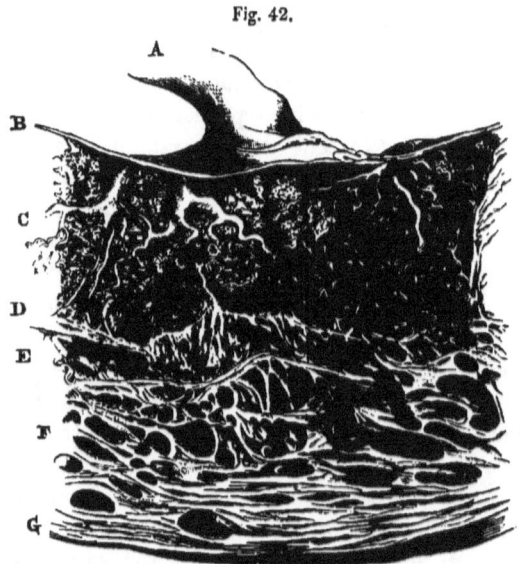

Fig. 42.—Section of the Human Uterus and Placenta at the thirtieth week of pregnancy. (After Ecker.)—A, umbilical cord; B, chorion; C, the foetal villi separated by processes of; D, cavernous *decidua*; E, F, G, wall of the uterus.

change; and, at parturition, the foetal villi are simply drawn out, like fingers from a glove, no vascular substance of the mother being thrown off.

The process by which the mucous membrane of the uterus returns to its unimpregnated condition after parturition in the pig has not been traced.

The extreme cases of placentation exhibited by Man and

by the Pig may be termed, with Von Baer and Eschricht, from the character of the maternal placenta, "caducous" and "non-caducous," or, from the degree of cohesion of the two placentæ in parturition, "coherent" and "incoherent;" or, what perhaps would be better still, the two Mammals may be spoken of as "deciduate" and "non-deciduate."* But, whatever terms be employed, the question for the classifier is to inquire what mammals correspond with Man and what with the Pig, and whether the groups of deciduate and non-deciduate *Monodelphia* thus formed are natural groups, or, in other words, contain such orders as can be shown, on other grounds, to be affined.

With respect to the *Deciduata*, it is certain that the Apes agree, in the main, with man in placental, as in other important characters; and, so far as has hitherto been observed (though our knowledge of the placentation of the Lemurs is very defective), their placentæ differ from those of Man only in presenting a more marked lobation—a character which occurs as a variety in Man.

The *Cheiroptera, Insectivora,* and *Rodentia* agree with Man in possessing a placenta which is not only as much "discoidal," allowance being made for the shorter curve of the uterine walls, as his, but also entirely resembles his in being developed in conjunction with a decidua. This *decidua* always corresponds to at least the *decidua serotina* of Man; frequently there is a well-developed *decidua reflexa*.† How far a *decidua vera* can be said to be developed is doubtful.

Figure 43 represents a section of the uterus, chorion, and partially-injected placenta of a Rat (the fœtus being one inch and a quarter long), taken in a direction perpendicular to the long axis of the uterine cornu. *a* is the mesometrium traversed by a large uterine vein; *b* is the wall of the uterus becoming looser in

---

* It is, of course, by no means intended to suggest by these terms, that the homologue of the *decidua* does not exist in the "non-deciduate" Mammals. The mucous membrane of the uterus becomes hypertrophied during pregnancy in both the deciduate and the non-deciduate Mammals; but it is thrown off, and so gives rise to a "decidua" only in the one of these two groups.

† See upon this subject the recently-published valuable essay of Reichert: "Beiträge zur Entwickelungs-geschichte des Meerschweinchens." Reichert finds a complete, or almost complete, *decidua reflexa* in Rats, Mice, Guinea-pigs, and Bats; while in Rabbits, Hares, and *Carnivora,* the *decidua reflexa* only partially surrounds the ovum.

texture and traversed by large venous channels in its inner substance, c; d is a decidual layer of the uterus of a cavernous structure, whence vascular processes are continued towards the chorionic surface of the placenta. A large vein (i) passes directly from the decidual layer (d), and the uterine sinuses beneath it, to near the chorionic surface of the placenta, beneath which it branches out horizontally. The chorion (f), rendered vascular over its non-placental part by the omphalo-meseraic

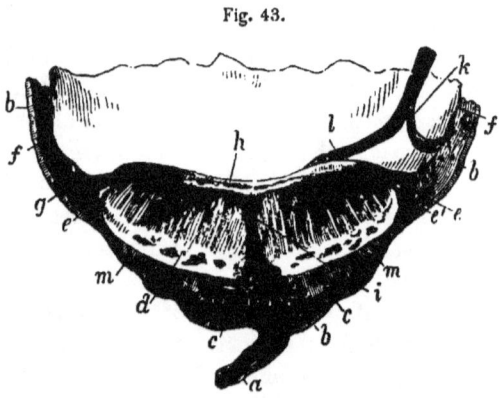

Fig. 43.—Magnified view of a section of the placenta and uterus of a pregnant Rat.

vessels (k) only begins to exhibit villous processes and folds at the point (g). These outermost villi appear to me to be free; but, more internally, they become closely connected with the upper surface of the placenta; and over the central third of the fœtal face of the placenta, the umbilical vessels (l) ramify in a radiating fashion, and send prolongations down between the decidual lamellæ. The slightest traction exerted upon the cord causes the placenta to separate along the line e, m, m, e, bringing with it, of course, the cup-shaped *decidua*, d.

The *Carnivora* develop a well-marked *decidua*, but their placenta in all genera which have been examined (except the Polecat, according to Von Baer) has the form of a complete zone, or broad girdle, surrounding the middle of the chorion and leaving the poles bare (Fig. 44).

Thus Man; the Apes, or so-called *Quadrumana*; the

Insectivora; the *Cheiroptera*; the *Rodentia*, to which the lowest apes present so many remarkable approximations; and the *Carnivora* (united into one group with the *Insectivora* by Cuvier) are all as closely connected by their placental structure as they are by their general affinities.

Fig. 44.

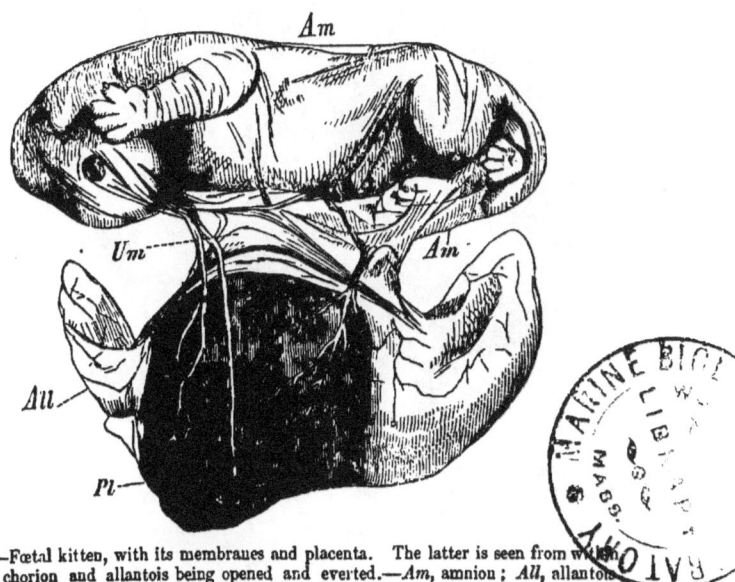

Fig. 44.—Fœtal kitten, with its membranes and placenta. The latter is seen from within, the chorion and allantois being opened and everted.—*Am*, amnion; *All*, allantois; *Pl*, placenta; *Um*, umbilical vesicle.
(From a preparation in the Museum of the Royal College of Surgeons.)

With the Pig, on the other hand, all the *Artiodactyla*, all the *Perissodactyla* (save one, taking the group in its ordinarily received sense), and all the *Cetacea* which have been studied, agree in developing no decidua, or, in other words, in the fact that no vascular maternal parts are thrown off during parturition. But considerable differences are observed in the details of the disposition of the fœtal villi, and of the parts of the uterus which receive them. Thus, in the Horse, Camel, and *Cetacea* the villi are scattered, as in the Pig, and the placenta is said to be *diffuse*; while in almost all true Ruminants, the fœtal villi are gathered into bunches, or cotyledons, which in the sheep (Fig. 46) are convex, and are received into cups of the mucous membrane of the uterus; while in the Cow, on the contrary, they are concave, and fit upon corresponding convexities of the uterus (Figs. 45 and 47).

Fig. 45.

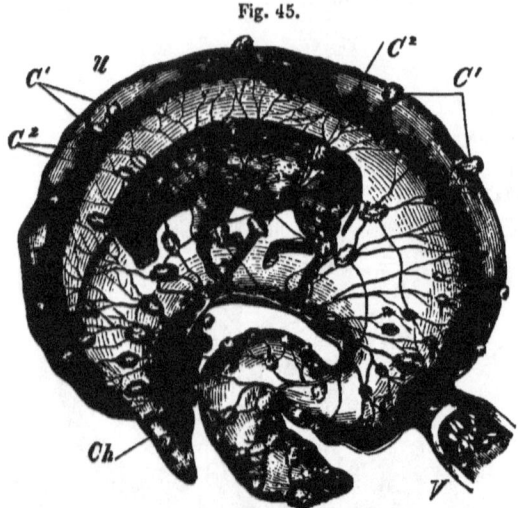

Fig. 45.—Uterus of a Cow in the middle of pregnancy laid open.— V, vagina; U, uterus; Ch, chorion; C^1, uterine cotyledons; C^2, fœtal cotyledons (after Colin).

Fig. 46.

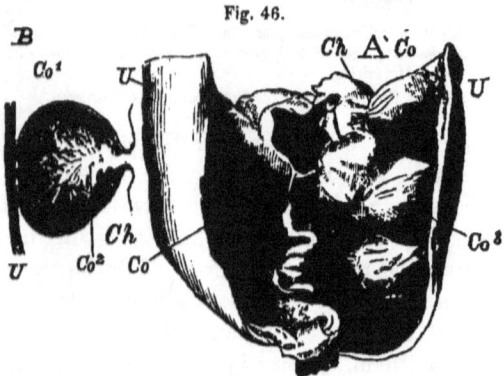

Fig. 46.—A. Horn of the Uterus of a pregnant Ewe, laid open to show, Ch, the chorion; with Co, the cotyledons.
B. Diagrammatic section of a Cotyledon.—U, uterine wall; Co^1, uterine cup of the cotyledon; Co^2, chorionic villous tuft of the cotyledon.
(From a preparation in the Museum of the Royal College of Surgeons.)

No one, probably, would be inclined to object to the association of the orders just mentioned into one great division of the

Fig. 47.—A fœtal cotyledon, C^2, half separated from the maternal cotyledon, C^1, of a Cow. *Ch*, chorion. *U*, uterus (after Colin).

Monodelphia, characterised by its placental structure. But such grouping leaves several important points for discussion. The Elephant has a zonary placenta,* and the genus *Hyrax* has been known, since the time of Home, to be in like case. Hence, as the elephants are commonly supposed to be closely allied with the *Pachydermata*, which possess diffuse, non-deciduate placentæ, and as *Hyrax* is now generally, if not universally, admitted into the same order as the Horse, which has a diffuse, non-deciduate placenta, it is argued that placental characters do not indicate natural affinities. A question, indeed, arises, which has not been answered by those who have described the placentæ of *Elephas* and *Hyrax*. Is the placenta of these animals simply a zone-like arrangement of villi or cotyledons, in connection with which no decidua is developed, or is it a true deciduate placenta, resembling that of the *Carnivora* in the essentials of its internal structure as in its external form? I have satisfied myself that, in both these animals, the placenta is as truly deciduate as that

* "Description of the Fœtal Membranes and Placenta of the Elephant," by Professor Owen.—*Philosophical Transactions*, 1857.

of a Rodent; so that most unquestionably, if the placental method of classification is to be adopted, both *Elephas* and *Hyrax* must go into the same primary division of the *Monodelphia* as the *Rodentia* and *Carnivora*.

But are these facts really opposed to the belief that the placenta has great taxonomic value?

So far as the elephants are concerned, I must confess that I see no difficulty in the way of an arrangement which unites the *Proboscidea* more closely with the *Rodentia* than with the *Artiodactyla* and *Perissodactyla*, the singular ties which ally the Elephants with the Rodents having been a matter of common remark since the days of Cuvier.

I have no hesitation in regarding *Hyrax* as the type of a distinct order of deciduate Monodelphous *Mammalia*. *Hyrax*, in fact, hangs by *Rhinoceros* mainly by the pattern of its molar teeth,—a character which affords anything but a safe guide to affinity in many cases.

The case of the *Edentata* presents greater difficulties. In this order, the Sloths have presented a cotyledonary placenta, and the Armadillos have been affirmed to possess a discoidal one. I am not aware that the minute structure of the placenta has been examined in either of these groups; but I am indebted to Dr. Sharpey for valuable information respecting the placental structure of *Manis*. The surface of the chorion is covered with fine reticulating ridges, interrupted here and there by round bald spots, giving it an alveolar aspect, something like the inside of the human gall-bladder, but finer. The inner surface of the uterus exhibits fine low ridges or villi, not reticulating quite so much. The chorion presents a band, free from villi, running longitudinally along its concavity, and there is a corresponding bald space on the surface of the uterus. The ridges of the chorion start from the margins of the bald stripe, and run round the ovum. The umbilical vesicle is fusiform. This is clearly a non-deciduate placenta, and the cotyledonary form of that of the Sloth leads me to entertain little doubt that it belongs to the same category. On the other hand, the placenta of *Orycteropus* is discoidal and deciduate.

It may also be urged that the value of the placenta as

an indication of affinity is weakened by the fact of a clear transition between the *Carnivora*, with a deciduate zonary placenta, and the *Cetacea*, with a non-deciduate diffuse placenta, being afforded by the extinct *Zeuglodon*.

But admitting all these difficulties and gaps in our information, it appears to me that the features of the placenta afford by far the best characters which have yet been proposed for classifying the Monodelphous *Mammalia*, especially if the concomitant modifications of the other fœtal appendages such as the allantois and yelk-sac, be taken into account.

Certain orders of the *Mammalia*, then have a DISCOIDAL DECIDUATE placenta. These are—

1. The PRIMATES—which have never more than four incisor teeth above, and as many below. The hallux is always provided with a flat nail (with occasional individual exceptions), and is capable of a considerable amount of abduction and adduction. All the *Primates* possess clavicles.

This order contains three sub-orders—

  a. The *Anthropidæ* (Man).—The dental formula is

$$i\frac{2.2}{2.2} \; c\frac{1.1}{1.1} \; pm\frac{2.2}{2.2} \; m\frac{3.3}{3.3}$$

and the teeth form a nearly even and uninterrupted series.

The hallux is nearly as long as the second toe, and is capable of comparatively slight movements of adduction and abduction. The arms are shorter than the legs, and, after birth, the latter grow faster than the rest of the body.

The habitual attitude of the body in standing or walking is erect.

  b. The *Simiadæ*.—(The Apes and Monkeys.)

The dental formula as in man, or with $pm \frac{3.3}{3.3}$.

The series of the teeth is uneven and interrupted by a diastema.

The hallux is considerably shorter than the second toe, and is capable of extensive adduction and abduction. The arms may be longer or shorter than the legs, but the latter do not grow faster than the rest of the body after birth.

  c. The *Lemuridæ*.—The dental formula varies. The

series of the teeth is uneven and interrupted. The hallux is large, while the second digit is always peculiarly modified, and sometimes rudimentary.

In this sub-order alone are the further characters found, that the lachrymal foramen is situated on the face; the orbit and temporal fossa communicate freely, as in the three succeeding orders; the clitoris of the female is perforated by the urethra; and more than one pair of teats may be present.

2. The INSECTIVORA.—The dentition varies. Usually, there are more than four incisors in each jaw, and the molars have sharp and pointed cusps.

The hallux possesses a claw, and has no marked freedom of adduction and abduction.

In all genera but one there are well-developed clavicles.

The so-called Flying Lemurs (*Galeopithecus*), the Hedgehogs, the Shrews, and the Moles belong to this order.

3. The CHEIROPTERA.—The dentition varies, but well-developed canines are always present.

The clavicles are strong and long. The fore-limbs are much longer than the hind limbs, and have the four ulnar digits prolonged. At least three of these are nailless. A prolongation of the integument, or "patagium," unites the prolonged digits together, and the arm with the body.

The hallux resembles the other digits. There are one or two pair of pectoral teats. The penis is pendent, and the testes remain in the abdomen.

To this order the Bats belong.

4. The RODENTIA.—There are never more than two incisors in the lower jaw, and usually only two, but sometimes four, in the upper. These incisors have persistent pulps, and continue to grow in adult life.

There are no canines. The molars vary in number from $\frac{2.2}{2.2}$ to $\frac{6.6}{5.5}$.

The hallux, when present, resembles the other digits, and the number of teats varies.

The Rats, Hares, Squirrels, &c., constitute this order.

Three orders possess a ZONARY DECIDUATE placenta—

1. The CARNIVORA.—The teeth are enamelled, and always consist of incisors, canines and molars. There are milk and permanent teeth.

The clavicles are always more or less rudimentary. The hallux and pollex, when present, resemble the other digits in mobility and in the character of their nails. The teats are not fewer than two pair, and are abdominal.

This order contains all the Cats, Hyænas, Civets, Dogs, Bears, Weasels, Racoons and Seals.

2. The PROBOSCIDEA.—The teeth consist only of tusk-like incisors, growing from persistent pulps, and molars. There is often only one set of teeth. The centra of the vertebræ are remarkably short for their breadth. There are no clavicles. The feet are five-toed, and the weight of the body is supported by a palmar and a plantar pad of integument, which underlies the toes. In standing, the knee-joint is straightened, and the femur lies in a line with the leg-bones, as in man.

The nose is prolonged into a proboscis.

The mammæ are two, and axillary.

The Elephants, Mastodons and *Dinotheria* constitute this order.

3. The HYRACOIDEA, containing the single genus *Hyrax*.—The dentition consists only of incisors and canines; and in the aged animal has the formula—

$$i \frac{2.2}{2.2} \ c \frac{0.0}{0.0} \ pm \frac{4.4}{4.4} \ m \frac{3.3}{3.3}.$$

The upper incisors have persistent pulps, and are long and curved, as in Rodents. The lower are straight, and grooved longitudinally. The molars resemble those of the Rhinoceros.

The front foot has four toes, the hind three. The inner nail of the hind foot is singularly curved.

There are no clavicles.

The stomach is simple. The intestine has two accessory cæca, in addition to the ordinary one. The ureters open into the fundus of the bladder, as in some Rodents. The penis of the male is pendulous, and the female has six teats, four inguinal and two axillary.

In the following orders the placenta is NON-DECIDUATE—

1. UNGULATA.—There are two sets of enamelled teeth. The molars have broad crowns with tuberculated or ridged surfaces.

There are never more than four full-sized toes on each limb, and the terminal digits are cased in thick hoof-like nails.

The members of this group are unguligrade or digitigrade, never plantigrade. Clavicles are never developed. The teats are few and inguinal, or numerous and abdominal, in position. The placenta is diffuse or cotyledonary.

The *Ungulata* are divisible into two sub-orders, which pass into one another: (*a*). the *Perissodactyla* (Horses, Rhinoceroses, Tapirs, *Palæotheria, Macraucheniæ*), with the third digit of each foot symmetrical in itself; the toes of the hind foot odd in number; a third trochanter on the femur; the stomach simple, and the cæcum very large; the horns, if present, median, and not supported by a bony core: (*b*). the *Artiodactyla* (Hippopotamuses, Pigs, *Anoplotheria*, Ruminants), with the third digit of each foot paired with the fourth, and the functional toes of the hind foot even in number; no third trochanter; the stomach more or less complex, and the cæcum not so large. The horns, if present, paired, and supported by a bony core.

The *Ungulata* are closely allied with the *Hyracoidea* among the Mammals with zonary placentation.

2. The CETACEA.—The body has a fish-like form, with a horizontal expansion of integument as a caudal fin, and sometimes a vertical expansion, as a dorsal fin. The hairs are very scanty.

The anterior limbs alone are developed, and are fin-like and devoid of nails. The nasal aperture, or apertures, are placed at the top of the head. There is no third eyelid, and the teats are two, and inguinal.

In the skeleton the cervical region is short, and the lumbar, long; there is no sacrum, and no odontoid process in the second cervical vertebra. The skull has a very broad brain case, and the premaxillæ, which are small in proportion to the maxillæ, are prolonged far in advance of the nasal aperture. The frontals have great supraorbital processes, and the maxillæ

THE ORDERS OF THE MAMMALIA. 103

extend on to, or over, them. The nasal bones are short or rudimentary, and the lachrymal bones are absent or small. The tympanic bone is thick and scroll-shaped. The mandible has a very small coronoid process, and the condyle is situated at the posterior extremity of the ramus.

There are no clavicles. There is no complete articulation between the bones of the fore-arm and the humerus, or between those of the carpus, fore-arm and digits. Some of the digits have more than three phalanges. The pelvis is rudimentary, and there is never more than a trace of hind-limb bones.

There are two sets of teeth only in the extinct *Zeuglodon*, but teeth may exist and be replaced before birth by baleen plates.

In this order the Whale-bone whales, the Dolphins, and the extinct Zeuglodonts are comprised. The *Cetacea* are closely allied with the *Carnivora*, among the mammals with zonary placentation.

Two orders of Monodelphous *Mammalia* remain. The placentation of one of them, the SIRENIA, is unknown.

Like the *Cetacea*, the *Sirenia* have a horizontally-flattened caudal integumentary fin, and the hind-limbs and sacrum are absent in existing genera, but in other respects they differ from the *Cetacea* completely, and approach the *Proboscidea* among the zonary and deciduate *Mammalia*.

They possess a fleshy snout and lips, a well-developed third eyelid, vesiculæ seminales, and salivary glands, all of which are absent in the *Cetacea*.

The teats are thoracic, and two in number.

The cervical region is short, but longer in proportion than in the *Cetacea*, though the number of the vertebræ may be only six (*Manatus*). There is an odontoid process. The heads of the ribs articulate with the centra of all the dorsal vertebræ, which is never the case in the *Cetacea*. The skull has an enormous zygomatic arch. The premaxillæ occupy a large space in the upper boundary of the gape of the mouth; and the mandible has a large coronoid and high ascending part of the ramus, in which respects it is opposed to that of the *Cetacea*. There are no clavicles. The bones of the fore-limb are freely articulated with one another, and the phalanges do not exceed three.

There is only one set of molar teeth, and horny plates are developed upon the premaxillary region of the palate and the opposed surface of the lower jaw. The apex of the heart is deeply bifid between the two ventricles.

The existing *Sirenia* are the estuarine, or littoral, Dugongs and Manatees.

In the remaining order, the EDENTATA, the placentation appears to vary, being diffuse and non-deciduate in *Manis*, cotyledonous (and non-deciduate ?) in *Bradypus*, and discoidal and deciduate in *Orycteropus*; but further investigation is needed before such variations can be safely admitted to exist.

The teeth are always devoid of enamel and of complete roots. There are never any median incisors in either jaw, and incisors are entirely absent in all but one genus of Armadillos.

The fore-limbs are well-developed, and their ungual phalanges are enveloped in long and strong claws. There are pectoral, and sometimes abdominal, mammæ.

The Sloths, the extinct *Megatherium* and its allies, the Anteaters, the Pangolins, and the Armadillos belong to this order.

The characters of the orders MARSUPIALIA and MONOTREMATA are the same as those of the sub-classes *Didelphia* and *Ornithodelphia*, of which they respectively constitute the sole members.

II. THE SAUROPSIDA.

The class AVES is divisible into three orders—

1. SAURURÆ.—The metacarpal bones are not ankylosed together. The caudal vertebræ are both numerous and large, so that the caudal region of the spine is longer than the body.

This order contains only the extinct bird, *Archæopteryx*.

2. RATITÆ.—The metacarpal bones are ankylosed together. The tail is shorter than the body.

The sternum is devoid of any crest and ossifies only from lateral and paired centres.

The long axes of the adjacent parts of the coracoid and scapula are parallel or identical. The barbs of the feathers are disconnected. The diaphragm is better developed than in other birds.

The Ostriches, Rheas, Emeus, Cassowaries, and the *Apteryx*, are the existing members of this order.

3. CARINATÆ.—The metacarpal bones are ankylosed, and the tail is shorter than the body, as in the *Ratitæ*, the terminal vertebræ being commonly ankylosed into a ploughshare-shaped bone.

The sternum possesses a crest or keel, and ossifies from a median centre in that keel, as well as from paired centres.

The long axes of the scapula and coracoid make an acute or slightly obtuse angle. The barbs of the feathers are usually connected.

To this order all ordinary birds belong.

The members of the class REPTILIA may be grouped into the following orders:—

1. The CROCODILA.—These reptiles have an epidermic exoskeleton consisting of horny scales, and a dermal exoskeleton of bony scutes, which may be confined to the dorsal surface of the body, or exist on the ventral aspect as well.

The centra of the dorsal vertebræ are procœlous or amphicœlous;* and in the middle and posterior dorsals a single transverse process supports both the capitulum and the tuberculum of the rib. Some of the ribs are provided with uncinate processes.

There are two sacral vertebræ. False ribs are developed as superficial ossifications in the wall of the abdomen.

The bones of the skull and face (except the mandible and hyoid) are solidly united together, and the presphenoidal region, which remains cartilaginous, is flattened laterally, so as to form an interorbital septum.

The nasal passages are shut off from the mouth by palatine plates of the maxillæ and palate bones, and (in modern Crocodiles) of the pterygoids as well. The tympanic cavities

* *Procœlous*, concave in front, and convex behind. *Amphicœlous*, concave on both faces.

are completely walled in by bone, and the Eustachian passages open on the base of the skull.

There are air passages connected with the tympana in the quadrate and articular bones, and in the supra-occipital.

The pterygoid unites only with the upper end of the quadrate; and the hyoidean apparatus is very simple, and not connected directly with the skull.

There are no clavicles. The pubes are greatly inclined forwards, and remain cartilaginous at their symphysial ends throughout life.

There are five digits in the fore-foot and four in the hind-foot, but only the three pre-axial (radial and tibial) digits bear nails.

The teeth are lodged in distinct alveoli, and are confined to the premaxillæ, maxillæ, and mandible.

The heart has four completely separated chambers, two auricles and two ventricles, but the right and left aortæ are connected by a small aperture immediately above their origin. The sclerotic is not ossified. The ear is provided with a moveable earlid; and the male has a grooved penis attached to the front wall of the cloaca, the aperture of which is longitudinal.

The modern Crocodiles, Alligators, and Caimans, and the extinct *Teleosauria* and Belodonts, form this order.

2. The LACERTILIA.—An epidermic and dermal exoskeleton is sometimes present, sometimes absent. The dorsal vertebræ have procœlous or amphicœlous centra; but their transverse processes are represented by simple tubercles, to which undivided proximal ends of the ribs are attached. There are two, or at most three, vertebræ in the sacrum. The presphenoidal region of the skull forms an interorbital septum.

The quadrate bone is usually movable on the skull, and the pterygoid is almost always connected with its distal end. The hyoidean apparatus is usually large and complicated. The limbs may be well developed, or one pair only present; or absent. A pectoral arch, consisting of clavicles and more or less ossified coracoscapular cartilages, is always present. The teeth are not lodged in sockets in any recent *Lacertilia*.

The heart has three chambers, two auricles and one ventricle, the cavity of the latter being partially divided, by a partition, into a right and a left portion.

There is a urinary bladder, and the aperture of the cloaca is transverse. The males have two eversible penes, one on each side of the cloaca.

The Lizards, the Blindworms, and the Chameleons, are the best known forms of this order.

3. The OPHIDIA.—The snakes have no dermal, or osseous, exoskeleton. The dorsal vertebræ are always proccelous, and have a rudimentary transverse process, with which the simple proximal ends of the ribs freely articulate. The front face of each vertebra gives off a wedge-shaped process (zygosphene), which fits into a corresponding pit (zygantrum) of the preceding vertebra.*

There is never any trace of a sternum or of a pectoral arch, of a fore limb, or of any sacrum; but in some few snakes (*Typhlops, Python, Tortrix*) there are rudimentary hind limbs.

There is no interorbital septum in the skull, and its lateral walls are completely osseous. The cartilaginous *trabeculæ cranii* remain distinct and persistent in the adult skull.

The quadrate bone is always more or less movable, and is generally united with the skull by the intermediation of the squamosal. The rami of the mandible are united at the symphysis only by ligament. The hyoidean apparatus is exceedingly rudimentary.

The teeth are never lodged in sockets, and are sometimes grooved or canaliculated on their front faces.

The heart is as in *Lacertilia*. The lungs and other paired viscera are usually unsymmetrical.

There is no urinary bladder. The copulatory organs are as in *Lacertilia*.

4. The CHELONIA.—The Turtles and Tortoises always possess an osseous exoskeleton, which becomes intimately united with parts of the endoskeleton to form a dorsal shield, the *carapace*, and a ventral shield, the *plastron*. To this are generally added epidermic horny plates, which form the so-called " tortoise-shell."

* Indications of this structure are found in some Lizards.

The dorsal vertebræ are immovably connected, and have no transverse processes, the proximal ends of the ribs uniting directly with the vertebræ. The spinous processes of most of the dorsal vertebræ expand into the median or "neural" plates of the carapace, while the ribs of these vertebræ enlarge into its lateral or "costal" plates. All these plates become united by sutures, so that the ribs are immovable. There are no sternal ribs, nor any sternum; the place of the latter being in part taken by the plastron, which usually consists of nine pieces, four pairs and one antero-median.

There are two vertebræ in the sacrum.

All the bones of the skull, except the mandible and the hyoid, are immovably united together.

The presphenoid cartilage forms an interorbital septum.

Both pair of limbs are well developed.

The jaws are ensheathed in a horny beak, and there are no teeth.

The heart is three-chambered, with a partial division of the ventricle, as in the two preceding orders. There is a large urinary bladder, and the males have a single grooved penis attached to the front wall of the cloaca.

These are all the orders of existing *Reptilia*, the following orders existing only in the fossil state—

1. The ICHTHYOSAURIA.—No exoskeleton is known. The centra of the vertebræ are short, broad and bi-concave, and the arches remain distinct from them throughout life. Transverse processes are represented only by small elevations of the centra. Those in the dorsal region are double, and articulate with the proximal ends of the ribs, which are deeply divided into distinct capitula and tubercula.

There is no sacrum. No sternal ribs, nor sternum, are known, but false abdominal ribs are developed as in the Crocodiles.

The skull has huge orbits, separated by an interorbital septum; and a long and tapering snout, formed chiefly by the premaxillæ. The nostrils are placed close to the orbits. There is a bony ring in the sclerotic.

The limbs are converted into paddles; the bones, from the humerus onwards, becoming broad and flattened, and losing their mobility on one another. The phalanges become very numerous, and marginal ossicles are added to the pre-axial and post-axial edges of the limb; but the number of the digits does not exceed five. There is a clavicular arch formed by an interclavicle and two clavicles. The pelvis is not directly connected with the vertebral column.

The teeth are lodged in grooves of the premaxillæ, maxillæ, and mandibles—not in distinct sockets. Their fangs are deeply folded.

Species of the genus *Ichthyosaurus* abounded during the mesozoic epoch.

2. The PLESIOSAURIA.—No exoskeleton is known. The centra of the vertebræ are flat, or slightly concave at each end, and the neural arches unite with the centra in the ordinary way. The dorsal vertebræ have long transverse processes, undivided at their ends, and articulating with the equally simple proximal ends of the ribs. No sternal ribs or sternum are known, but there are well-developed false, or exoskeletal, abdominal ribs.

There is a sacrum composed of two vertebræ, and the cervical region is often extremely long.

The snout is produced, and the external nostrils placed far back near the large orbits.

There is no bony ring in the sclerotic.

The limbs are paddle-like, but the bones retain the normal form much more than in the case of the *Ichthyosauria*. There are no marginal ossicles. A clavicular arch, formed of clavicles and interclavicle, seems to have existed in some species, if not in all.

The scapula has a remarkable form, and sends out a preglenoidal process, as in the *Chelonia*.

The genera *Plesiosaurus, Simosaurus, Nothosaurus*, which constitute this group, are mesozoic marine reptiles.

3. The DICYNODONTIA.—These strange reptiles (*Dicynodon, Oudenodon*) are known only from strata of triassic (?) age which occur in both Africa and India. They are not known to possess

any exoskeleton. Their dorsal vertebræ are amphicœlous, with crocodilian transverse processes.

The sacrum is large, and formed by the ankylosis of many, sometimes six, vertebræ.

The cranial and upper facial bones are all firmly fixed together, in which respect, and in the conversion of the jaws into a kind of beak, which appear to have been sheathed in horn, they resemble the Chelonia.

Teeth seem to have been absent in some species, in others there was a pair of great tusks, with persistent growth implanted in the upper jaw.

The pectoral and pelvic arches were very strong, and the limbs well developed and fitted to support the massive body on land.

4. The PTEROSAURIA.—The flying Lizards of the Mesozoic epoch are not known to have possessed any exoskeleton.

The dorsal vertebræ are procœlous, with crocodilian transverse processes: ossified sternal ribs, and splint-like, false, or exoskeletal, abdominal ribs are present.

There is a broad sternum, with a median keel or crest.

The skull is in many respects very bird-like, but the jaws carried teeth, implanted in alveoli. In some genera the extremities of the jaws are edentulous, and seem to have been sheathed with horn.

The sclerotic has an osseous ring.

The pectoral arch is extremely like that of a carinate bird, but no clavicles have been discovered. The manus has four digits, three of which are short and provided with claws, while the fourth, enormously prolonged and clawless, appears to have supported a "patagium," as in the Bats.

The posterior limbs are comparatively small.

The long bones and the vertebræ appear to have contained pneumatic cavities, as in many birds.

5. The DINOSAURIA.—The bony exoskeleton is sometimes more highly developed than in the *Crocodilia*, and sometimes absent. The centra of the posterior dorsal vertebræ are flat or slightly concave at each end, and they have crocodilian transverse processes and ribs. The centra of the anterior dorsal and

of the cervical vertebræ are sometimes concave behind and convex in front (*opisthocœlous*). There are four, or more, vertebræ in the sacrum.

The pelvis and bones of the hind limb are in many respects very like those of birds. No clavicles have been observed, and the fore limb is sometimes very small in proportion to the hind limb.*

III. THE ICHTHYOPSIDA.

The AMPHIBIA are divisible into four orders—

1. The URODELA.—There is no exoskeleton. The dorsal vertebræ are amphicœlous or opisthocœlous, and have single or bifid transverse processes, to which short ribs of a corresponding form are attached. The single sacral vertebra has movable ribs like the rest, when a sacrum exists. The caudal vertebræ are numerous and distinct. The bones of the fore-arm and of the leg remain separate, and the proximal bones of the tarsus are not elongated.

This order comprises the Newts and Salamanders, with the so-called "perennibranchiate" *Amphibia*, such as the *Proteus, Siren*, &c.

2. The BATRACHIA.—An exoskeleton is rarely represented by dermal ossifications in the dorsal region of the trunk. The dorsal vertebræ are procœlous, and have simple and long transverse processes, with only rudimentary ribs. The single sacral vertebra has wide lateral processes for articulation with the ilia, and no movable ribs. A styliform ossification takes the place of the centra of the caudal vertebræ.

The ischia and pubes of opposite sides are applied together, and unite by their inner faces. The radius and ulna in the fore limb, and the tibia and fibula in the hinder extremity, unite into a single bone. The calcaneum and astragalus are greatly elongated.

The Frogs and Toads compose this order.

* In *Iguanodon, Megalosaurus, Poikilopleuron*, and *Scelidosaurus*, the distal end of the tibia extends outwards behind the fibula in a manner which is extremely peculiar and characteristic; and the astragalus is very like that of a bird.

3. The GYMNOPHIONA have rounded, worm-like bodies, which are devoid of limbs and tail. They have scales imbedded in the integument. The dorsal vertebræ are bi-concave, and possess double transverse processes, with which the capitula and tubercula of the ribs articulate.

The genera *Cæcilia, Siphonops, Ichthyophis,* and *Rhinatrema* belong to this order.

4. The LABYRINTHODONTA.—The body is salamandriform, with relatively weak limbs, and a long tail. The dorsal vertebræ, when completely ossified, are bi-concave, with double transverse processes. The ribs have distinct capitula and tubercula.

In the thoracic region, three superficially sculptured exoskeletal plates, one median and two lateral, occupy the place of the interclavicle and clavicles. Between these and the pelvis is a peculiar armour, formed of rows of oval dermal plates, which lie on each side of the middle line of the abdomen, and are directed obliquely forwards and inwards, to meet in that line.

The skull has distinctly ossified epiotic bones in the same position and of the same form as those of fishes. The cranial bones are sculptured, and many exhibit peculiar smooth, symmetrical grooves—the so-called "mucous canals."

The parietes of the teeth are deeply plaited and folded, so as to give rise to a complicated "labyrinthine" pattern in the transverse section of the tooth.

Remains of the Labyrinthodonts, which sometimes attained a large size, are found from the Carboniferous to the Triassic or Liassic strata, inclusively.

The class PISCES is divisible into six orders—

1. DIPNOI.—There is a skull with distinct cranial bones, and a mandible. The notochord is persistent, and there are no vertebral centra.

There are two pairs of filiform fins, each supported by a single, jointed, cartilaginous rod. The posterior pair are placed close to the anus. The pectoral arch has a clavicle.

The heart has two auricles, and true lungs coexist with

rudimentary external branchiæ and functional, pectinated, internal branchiæ.

This order contains only the "Mud-fish," *Lepidosiren* and *Rhinocryptis*.

2. The ELASMOBRANCHII.—The skull and mandible are well developed, but there are no cranial bones.

The condition of the vertebral column varies.

There are two pair of fins, each supported by many series of cartilages. The posterior pair are placed close to the anus.

The pectoral arch has no clavicle.

The heart has one auricle, and a rhythmically contractile bulbus arteriosus, which contains striated muscular fibre in its walls, and is provided with several transverse rows of valves. The gills are pouch-like. The optic nerves form a chiasma.

To this order belong the Sharks, Rays, and *Chimœræ*.

3. The GANOIDEI.—The skull has cranial bones, and there is a mandible. The condition of the vertebral column varies. There are usually two pair of limbs of the same essential structure as those of the *Elasmobranchii*. The pectoral arch has a clavicle, and the posterior limbs are placed close to the anus.

The heart and optic nerves are as in the *Elasmobranchii*.

The gills and the opercular apparatus are as in the *Teleostei*. This order includes the Sturgeons and bony Pikes (*Lepidosteus*), besides *Polypterus*, *Amia*, and numerous extinct genera.

4. The TELEOSTEI comprises the majority of existing genera of fishes provided with a bony skeleton. The skull has cranial bones and a mandible. More or less ossified distinct vertebræ are always present. The limbs, when they exist, have the same general structure as those of the Ganoids, but the ventral fins vary in position. The pectoral arch has a clavicle.

The gills are pectinated or tufted; a bony operculum and pre-operculum and branchiostegal rays are always developed.

The bulbus aortæ is not rhythmically contractile, and it is separated from the ventricle by only a single row of valves.

The optic nerves cross, but do not form a chiasma.

5. The MARSIPOBRANCHII includes the Lampreys and Hags.

There is a cartilaginous skull, devoid of cranial bones; but no mandible, nor any traces of limbs. The notochord is persistent and no centra of vertebræ are developed.

The heart presents an auricle and a ventricle. The gills are sac-like, and not ciliated.

These fishes have a distinct brain and auditory organs; and the liver and kidneys are constructed upon the ordinary vertebrate plan.

6. The PHARYNGOBRANCHII.—Neither cartilaginous nor osseous skull, nor mandible, nor any limbs, are developed. The persistent notochord extends beyond the cerebro-spinal axis to the anterior end of the body, whereas in all other *Vertebrata* it stops behind the pituitary gland.

There are no vertebral centra or arches. No distinct brain exists, nor any auditory organs.

There is no heart, but several of the great vessels are rhythmically contractile. The walls of the pharynx are perforated by many slits, and ciliated. The liver is sac-like, and no kidneys have been discovered.

This order is represented by the single species *Amphioxus lanceolatus*.

CHAPTER VI.

THE ORDERS INTO WHICH THE CLASSES OF THE INVERTEBRATA ARE DIVISIBLE.

I. THE MOLLUSCA.

THE CEPHALOPODA are divided into two orders, the *Dibranchiata* and the *Tetrabranchiata*.

1. In the DIBRANCHIATA, the processes ("Arms") into which the margins of the foot are divided are not more than ten in number, and they are provided with acetabula, or suckers.

The funnel is a complete tube. When a pallial shell is developed, it is internal, or enveloped by a prolongation of the mantle, and does not lodge the body. The beaks are horny. There are only two gills; and there is an ink-bag.

2. In the TETRABRANCHIATA, represented by the Pearly Nautilus, the processes into which the margins of the foot are divided are very much more numerous. They bear no acetabula, but each contains a retractile tentacle. The funnel is open below. The shell is pallial, external, chambered, and siphunculated. The beaks are more or less covered with calcareous matter. There are four gills, and no ink bag.

I am not satisfied that any good divisions having the value of orders have at present been formed among the *Pteropoda*, *Pulmogasteropoda*, *Branchiogasteropoda*, and *Lamellibranchiata*.

II. THE MOLLUSCOIDA.

The ASCIDIOIDA are divisible into three orders—

1. The BRANCHIALIA, in which the branchial sac is very large, in proportion to the intestine and generative viscera, so that these usually lie on one side of it.

This order contains the "Solitary Ascidians," *Perophora*, and the *Botryllidæ, Pyrosomidæ,* and *Salpidæ.*

2. The ABDOMINALIA, in which the branchial sac occupies a comparatively small proportion of the body and lies altogether in front of the intestine and reproductive organs. *Clavelina, Amoroucium,* and the remaining "Compound Ascidians."

I formerly proposed a third order—the *Larvalia*—to contain *Appendicularia,* which differs from all the rest in retaining the larval tail as a locomotive organ, and in many other peculiarities. But as, up to this time, all the individuals of this genus which have been discovered have been males, it is possible that the females will turn out to be more complete and of a more ordinary type of structure.

The BRACHIOPODA.—Of these there are two well-marked orders—

1. The ARTICULATA.—The valves of the shell are connected along a hinge-line, which is often provided with teeth and sockets. The lobes of the mantle are united upon the dorsal side of the body. The intestine ends in a blind sac.

The *Terebratulidæ, Rhynchonellidæ, Spiriferidæ,* and *Orthidæ,* belong to this order.

2. The INARTICULATA.—The valves of the shell are not connected along a hinge-line. The lobes of the mantle are completely separated. The intestine terminates in an anus on one side of the body. This order contains the *Craniadæ, Discinidæ,* and *Lingulidæ.*

The POLYZOA are divisible into two orders—

1. The PHYLACTOLŒMATA.—The lophophore, or tentaculated oral disk, is bilaterally symmetrical, and a process—the epistome—overhangs the mouth.

The *Lophophea* and *Pedicellinea* are comprised in this order.

2. The GYMNOLŒMATA have no epistome, and the lophophore is orbicular.

This order comprehends the *Urnatellea, Paludicellea, Cyclostomata, Ctenostomata,* and *Cheilostomata.*

III. THE CŒLENTERATA.

The ACTINOZOA.—I cannot distinguish more than two groups of ordinal value in this class.

1. The CTENOPHORA.—Free swimming organisms, provided with symmetrically disposed bands of large cilia which subserve locomotion. At the end of the body opposite the mouth the common cavity opens externally—in some species by two apertures between which a nervous ganglion and otolithic sac may be discovered. The mesenteries are very thick, while the intermesenteric spaces take the form of canals. The genera *Beröe, Cydippe, Cestum*, &c., belong to this order.

2. The CORALLIGENA.—These animals are organized upon the same plan as the *Actinia* described above, with variations in the number and proportion of the parts, and in the forms of the masses which are produced by the gemmation or fission of the first-formed individual. Most of them give rise to a calcareous skeleton, which may have the form of detached spicula, of a solid axis, or of a *theca* or cup for each actiniform zoöid, or of some combination of these.

The *Actiniæ, Gorgoniæ*, and coral-forming animals in general constitute this order.

The HYDROZOA.—Three divisions of ordinal value may be distinguished among the animals constituting this extensive class.

1. The HYDROPHORA.—The alimentary zoöid, or polype, is provided with numerous tentacles, which are either set round its mouth or scattered over its surface. When free-swimming reproductive zoöids are developed, the genitalia are borne by a *nectocalyx*, or swimming-bell, the inner margin of the aperture of which is produced into a muscular velum. Such zoöids are always formed by budding, and are commonly called craspedote, or gymnophthalmous, *Medusæ*.

The immediate product of the growth of the embryo is a fixed Hydroid, Tubularian or Sertularian Polype.

2. The SIPHONOPHORA.—The alimentary zoöid never bears numerous tentacles around its mouth or on its surface, but such

organs are developed either singly from its base, or arise from the common body. When free-swimming reproductive zoöids are developed, they are craspedote *Medusæ*.

The immediate product of the growth of the embryo is a free Calycophoridan or Physophoridan Polype.

3. The DISCOPHORA.—The immediate product of the growth of the embryo is a fixed polype provided with tentacles arranged on the circumference of the body, at some distance from the mouth. When free-swimming reproductive zoöids are produced, they are developed by transverse fission from this primary form, and consist of a disk or umbrella, devoid of any velum, from the centre of which the gastric apparatus depends. This is called an acraspedote, or steganophthalmous, *Medusa*.

Lucernaria, Cyanæa, Rhizostoma, and their allies constitute this order.

Not improbably a fourth order will have to be added to these three, for the *Trachymedusæ* of Haeckel.

IV. THE ANNULOSA.

The arrangement of the class INSECTA into orders is hardly to be regarded as complete at present, but several groups are very clearly distinguishable from one another. These are—

1. The COLEOPTERA, which are provided with mandibles and maxillæ adapted for masticatory purposes. The wings are rarely absent; when present the anterior pair are converted into horny or leathery *elytra*, which are not used in flight, and serve during rest as covers to the posterior wings. These are membranous and can be folded up longitudinally and transversely.

Metamorphosis is complete; a quiescent pupa stage, in which the limbs of the pupa are free, being interposed between the active larva and the sexual, or imago, state.

This order contains the Beetles, Cockchafers, Weevils, etc.

2. The HYMENOPTERA have mandibles and maxillæ adapted for biting and cutting. The first pair of maxillæ may be produced into knife-like blades, and the second pair into a sucking proboscis. Both pair of wings take part in flight, and are alike membranous, transparent, and reticulated. In the females

certain appendages of the hinder somites of the abdomen become converted into a sting for offence and defence, or a borer or saw for oviposition.

Metamorphosis is complete; the limbs of the pupa are free.

The Bees, Wasps, Ants, Ichneumon-flies, and Saw-flies belong to this order.

3. The LEPIDOPTERA.—The labrum and mandibles are aborted, and of the second pair of maxillæ (or labium), only the palpi are well developed. The first pair of maxillæ are prolonged, and give rise by their union to a tubular suctorial proboscis.

Wings are absent only in the females of a few genera. When present, both pairs are reticulated and membranous in texture, and are covered with delicate cuticular scales.

Metamorphosis is complete. The larvæ have masticatory jaws, and spurious feet attached to several of the somites of the abdomen. The limbs of the pupæ are firmly fixed to the sides of the body by its outer coat.

The Butterflies and Moths compose this order.

4. The DIPTERA.—The labium (the coalesced second pair of maxillæ) is produced and forms the chief part of a proboscis. The mandibles and maxillæ are converted into styliform cutting organs connected with this. The maxillary palpi are well developed, the labial palpi absent.

Only the anterior pair of wings is well developed and used in flight. They are membranous and naked; the posterior wings are converted into little knobbed stalks, the *halteres*.

Metamorphosis is complete. The larvæ are apodal grubs. The pupæ are either inclosed in the hardened larval skin, or resemble those of the *Lepidoptera*. In the latter case, when the pupæ are aquatic in habit, they may swim about by using the abdomen as a locomotive organ.

This order contains the Flies, Gnats, and Fleas (the last are without wings).

5. The HEMIPTERA.—The labium is produced and divided into three or four joints. The labrum, mandibles and maxillæ are more or less produced into styliform cutting organs. Neither labial nor maxillary palpi are present.

The anterior wings may be more or less dissimilar to the posterior, and simulate elytra.

Metamorphosis is incomplete; that is, the insect remains active through all its changes, from the larval to the perfect condition, except in the case of the male *Coccidæ*, which undergo a complete metamorphosis.

The Bugs, Cicadas, Lantern-flies, Plant-lice, Cochineal Insects and Lice belong to this order.

The proper grouping of the insects which do not fall into any of the orders which have now been mentioned is still a matter of doubt.

Those with a complete metamorphosis have been distinguished under the following heads—

1. STREPSIPTERA.—The jaws are abortive and useless for mastication. The maxillary palpi are present, while those of the labium are absent.

The anterior wings are represented by twisted rudimentary appendages of the mesothorax. The posterior wings are large, folding longitudinally.

The larvæ, which are parasitic on the bodies of *Hymenoptera*, are at first six-legged and active, but lose their limbs and become vermiform.

The females remain in this condition throughout life, while the males undergo a complete metamorphosis and are extremely active in the imago state.

2. TRICHOPTERA.—The jaws are abortive, but both maxillary and labial palpi are developed.

The wings are dissimilar, hairy or scaly; the posterior wings are usually folded.

The larvæ resemble those of the *Lepidoptera*. They have masticatory jaws and are aquatic, constructing cases in which they reside and eventually undergo their metamorphosis. They commonly breathe by tracheal gills.

The Caddis-flies exemplify this group.

3. NEUROPTERA.—The jaws are well developed and masticatory; both maxillæ and labium are provided with palpi. The wings are similar and membranous.

THE INSECTA.

Metamorphosis is complete; a quiescent pupal condition being interposed between the larval and the imaginal states.

The Ant-lions, Lace-flies, and Scorpion-flies are included under this division.

Of the remaining insects, which have an incomplete metamorphosis and masticatory jaws, many have been included under the head of

4. ORTHOPTERA.—But the assemblage is a very miscellaneous one, and contains a number of remarkably different types, of which the chief are—

a. The restricted *Orthoptera*, which have the anterior wings coriaceous and elytra-like, the posterior wings membranous and folded.

These are the Cockroaches, Mantides, Leaf and Stick Insects, Grasshoppers and Locusts.

b. The *Dermatoptera.*—The anterior wings are converted into elytra; the posterior membranous and doubly folded.

The Earwings alone are contained in this group.

c. The *Corrodentia*, or *Termitinæ*, have the fore and the hind wings similar and not folded.

The genus *Termes*, which belongs to this division, includes the so-called "White Ants," whose destructive ravages are so well known in tropical countries. Besides sexual individuals, there are two sexless forms, termed "soldiers" and "workers." The male and female Termites alone have wings. The hinder wings are not folded, and both pairs fall off after copulation.

d. The *Perlariæ* have membranous hairy wings, the posterior folded lengthwise; and aquatic larvæ, with more or less distinct tracheal gills.

e. The *Ephemeridæ*, or Day-flies, have the jaws rudimentary; the posterior wings not folded, much smaller than the anterior, or absent.

The larvæ are carnivorous, with well-developed jaws and tracheal gills. The imagines undergo ecdysis after leaving the pupa, in which circumstance they stand alone among insects.

f. The *Libellulidæ*, or Dragon-flies, have well-developed jaws and are predaceous carnivores. The wings are similar, membranous and reticulated.

The larvæ are provided with tracheal gills, and are as predaceous as the imagines; the labrum being converted into a peculiar mask-like apparatus.

The wings of the *Libellulidæ* are moved by muscles which are attached directly to them; and the males have a peculiar copulatory apparatus attached to the ventral portion of the second somite of the abdomen.

Three groups of insects, with incomplete metamorphosis, remain, which do not fit well into any of the preceding assemblages—

a. The *Physopoda*, comprising the genus *Thrips* and its allies, have a proboscidiform mouth, styliform mandibles, and maxillæ united with the upper lip. Both maxillary and labial palpi are developed; the wings are similar and unfolded.

b. The *Thysanura* have biting jaws, no wings, and scaly bodies. Some possess a peculiar springing apparatus developed from the abdomen.

The genera *Lepisma* and *Podura* belong to this division.

c. The *Mallophaga*, or Bird-lice, have masticatory jaws, are devoid of wings, and live parasitically, chiefly upon birds, whose feathers they devour.

The MYRIAPODA are divided into two orders—

1. The CHILOPODA (Centipedes) have the head and the segments of the body broad and depressed, and each segment bears only one pair of limbs, which are separated by a broad sternum.

The aperture of the organs of reproduction is situated at the posterior extremity of the body, and the males have no copulatory organs.

2. The CHILOGNATHA (Millipedes) have the head and segments of the body rounded or compressed. All but a few of the most anterior segments bear two pairs of limbs, attached close together upon each side of the middle line.

The reproductive aperture and the copulatory organs of the males are situated on the ventral surface of the second to the seventh segments of the body.

The ARACHNIDA are divisible into six orders—

1. The ARTHROGASTRA have the abdomen distinctly divided into somites, and passing by a broad base into the cephalothorax. Respiration takes place in some by lung sacs, and in some by tracheæ.

Scorpio, Chelifer, Phrynus, Phalangium, Galeodes, are the principal genera embraced by this division.

2. The ARANEINA (or Spiders) have the abdomen not segmented, and connected with the cephalothorax by a narrow peduncle. The antennæ are subchelate. The mandibular palpi are filiform, their extremities being peculiarly modified in the males. The two pair of maxillæ are leg-like.

Four or six conical papillæ are situated on the ventral surface of the abdomen, in front of the anus, and give exit to the secretion of the silk glands.

There are two or four pulmonary sacs, and two stigmata connected with tracheæ.

3. The ACARINA (or Mites and Ticks) have the abdomen unsegmented, and passing without any constriction into the cephalothorax.

The parts of the mouth are frequently converted into a suctorial apparatus. When distinct respiratory organs are present they take the form of tracheæ.

The foregoing are what may be called the typical *Arachnida;* the three following orders are aberrant forms—

1. The fresh-water ARCTISCA or TARDIGRADA (Water-bears), are minute animals, which have no distinct abdomen, the vermiform body representing the cephalothorax, which possesses four pair of rudimentary limbs. The fourth pair are completely posterior.

The mouth is a suctorial tube, containing two stillets. There are no respiratory or circulatory organs, and the sexual apparatus is hermaphrodite.

2. The PYCNOGONIDA are marine, and also have the abdomen rudimentary; but the legs, of which there are four pair, are enormously long, and many-jointed. There is a chambered heart, but no respiratory organs. The stomach sends cæcal prolongations into the legs. The sexes are distinct.

3. The PENTASTOMIDA have long, annulated, vermiform bodies. Two hooks, representing limbs, are placed upon each side of the mouth at the anterior end of the body. There are no respiratory or circulatory organs, and the sexes are distinct.

In the embryos, the anterior part of the body is provided with four short, articulated limbs; while, posteriorly, it is produced into a relatively short abdomen.

These are parasitic animals. *Pentastoma tænioides*, which lives in the frontal sinuses of the dog, is the sexual form of *P. denticulatum*, which is found in the lungs and liver of Rabbits.

The Orders of the CRUSTACEA are—

1. The PODOPHTHALMIA.—These have twenty somites in the body; a carapace developed from more or fewer of the thoracic somites; branchiæ attached to the thoracic limbs or somites, and a short more or less quadrate heart. With rare exceptions, the eyes are placed upon movable footstalks.

This order contains the Crabs, Lobsters, Shrimps, and *Diastylidæ*.

2. In the BRANCHIOPODA, the body is composed of more or fewer than twenty somites. The thoracico-abdominal appendages are foliaceous, and resemble the anterior maxillipeds of the *Podophthalmia*. The carapace, if present, is altogether cephalic. The heart varies.

Nebalia, Apus, Branchipus, Daphnia, belong to this order.

3. The OSTRACODA possess a hard, bivalve, hinged carapace. The total number of appendages, including those attached to the head, does not exceed seven pair, and the thoracic limbs are strong legs. The antennæ are similar in form and function to the latter. The abdomen is altogether rudimentary, and there is no heart. To this order belong the genera *Cypris* and *Cythere*.

4. The PECTOSTRACA are *Crustacea* which are fixed in the

adult condition, but leave the egg as actively locomotive larvæ, provided with a discoid carapace and three pair of appendages, which represent the antennules, antennæ, and mandibles of the higher *Crustacea*. After a time, these larvæ moult, the carapace becomes bivalved, the antennules are converted into organs of prehension, and several pair of paddle-like thoracic limbs appear. The further state of the larvæ differs in the two principal subdivisions of this order.

In the *Rhizocephala*, which are parasitic upon other Crustacea, the anterior region of the body gives out filaments, which grow like roots into the body of the animal which bears the parasite.

In the *Cirripedia* the larvæ first adhere to foreign bodies by the suckers developed upon the prehensile antennæ, and then glue themselves in this position by the secretion of a " cement gland." The anterior part of the body next becomes enlarged either in breadth or length, or both, into a peduncle; and, usually, definite calcifications occur in the carapace. The thoracic limbs of the larva are commonly preserved as six pair of cirrated appendages, and the animals are very generally hermaphrodite. Sometimes, however, they are diœcious; and some of the hermaphrodite forms have "supplemental males."

The " Acorn Shells " and Barnacles are common examples of *Cirripedia*.

5. In the STOMAPODA the axis of the ophthalmic and antennulary somites is parallel with that of the body generally, and these two somites are ordinarily movable on one another. There is a carapace developed from the head. The eyes are on movable stalks. The heart, elongated and many-chambered, extends through the abdomen. The branchiæ are attached to the abdominal appendages.

To this order the *Squillæ* belong.

6. The EDRIOPHTHALMIA have sessile eyes. At fewest, five of the posterior thoracic somites are free and movable, and the anterior pair of thoracic appendages are converted into foot-jaws. There is no carapace. The heart is many-chambered, and the branchiæ may be thoracic or abdominal.

Most of these are aquatic and marine animals; the Woodlice are their terrestrial representatives.

7. The MEROSTOMATA.—The body is divided into an anterior shield-shaped portion, which bears the sessile eyes on its upper surface; a middle part is formed by the rest of the somites of the body, and there is a terminal broad, or spine-like, telson, In the living King Crab (*Limulus*) there are thirteen pair of appendages. The five anterior pair belong to the head, but all have the form of locomotive limbs. The five posterior pair are broad lamellæ, united in the middle line, on the posterior faces of which the gills are borne. The heart is elongated and eight-chambered. In the extinct *Eurypterida* only the head is provided with distinct appendages, and, of these, the posterior pair are styliform, or paddle-like, swimming organs.

8. The COPEPODA are minute aquatic crustaceans, which inhabit both the sea and fresh-water. Many, like the common *Cyclops*, are actively locomotive animals, while many, on the other hand, are among the most sluggish and strangely-modified of fixed parasites. These last have been termed *Epizoa*.

The locomotive forms have not more than eighteen somites in the body, the anterior part of which is covered by a carapace provided with a single, or double, median eye upon its dorsal surface. The antennules are large oar-like appendages, and are the chief agents in locomotion, which is aided by the paddle-like posterior thoracic members.

The heart is short, and there are no distinct respiratory organs. The sexes are in distinct individuals; the female, among the parasitic forms, being often very much larger than the male. The larvæ leave the egg as oval bodies, provided with two or three pair of oar-like limbs.

9. The TRILOBITA are Crustacea which, like the *Eurypterida*, have been extinct since the palæozoic epoch. The form of the head is as in the *Merostomata*; in most cases it bears sessile eyes of a similar character. On its under surface a large labrum is attached, but no jaws or other appendages have yet been discovered, and the ventral wall of the body seems to have been completely membranous. The cephalic shield is dis-

tinguished by the presence of a suture dividing its median from its lateral regions.

The ANNELIDA are primarily divisible into two groups—the *Chætophora* and the *Discophora*. The former has probably more than a merely ordinal value.

1. The CHÆTOPHORA have chitinous setæ developed in sacs of the integument. These are sometimes merely implanted in the walls of the somites, at others, they are borne by distinct rudiments of limbs, or parapodia. They never possess suckerlike organs of adhesion.

The *Errantia* and *Tubicola*, the Earthworms, and the *Gephyrea* form this division.

2. The DISCOPHORA (or Leeches) never possess parapodia nor lateral series of setæ occupying their place, but they are provided with one or more sucking disks.

V. THE ANNULOIDA.

The groups comprehended under the title of the SCOLECIDA are probably of greater importance than orders, but for the present I shall take them as such.

1. The TREMATODA (or Flukes) possess an alimentary canal, but are always devoid of an anus, and are usually provided with one or more suckers for adhesion to the animals in, or on, which they are parasitic. The body is not covered with cilia in the adult state, nor is it segmented. There is only one set of sexual organs—usually hermaphrodite. The larvæ are often tailed or cercariform—never provided with three pair of hooks. The species have asexual conditions, but are never "cystic."

2. The TÆNIADA (or Tape Worms and Bladder Worms) possess no alimentary canal. The anterior end of the body is provided with suckers, hooks, or foliaceous appendages, or with a combination of all three. The hermaphrodite reproductive organs are serially repeated, many times, in the body. The larvæ are oval, and provided with three pairs of hooks, arranged in bilateral symmetry. Beside the "cestoid" sexual forms,

there are "cystic" asexual conditions of most, if not all, the species. The *Tæniada* are internal parasites.

3. The TURBELLARIA are non-parasitic. Their flattened or elongated bodies are covered with cilia throughout life, and are not provided with suckers. They may be hermaphrodite or unisexual. To this order belong the genera *Planaria* and *Nemertes*.

4. The ACANTHOCEPHALA are unsegmented, vermiform, internal parasites (*Echinorhynchus*), devoid of any alimentary canal, and adhering by means of a proboscis, which is covered with spines. The sexes are distinct, and there is only one set of reproductive organs in each individual.

5. The NEMATOIDEA (or Thread Worms, including the *Gordiacea*) have vermiform non-ciliated bodies, devoid of segmentation. The integument is dense, and beneath it lies a single layer of muscles, usually divided into four longitudinal bands. The intestine is straight, extending from the mouth, at the anterior end of the body, to the anus, which is more or less near the posterior end. In some few species the intestine is abortive. The sexual organs are tubular glands, opening externally by a single aperture, and are usually lodged in distinct individuals. The embryo at once takes on the form of a worm.

There are both free and parasitic Nematoids, and some species pass part of their lives in the one condition, and part in the other.

6. The ROTIFERA (or Wheel-animalcules) have subcylindrical bodies, more or less distinctly annulated. The oral end is provided with cilia, arranged to form the chief locomotive organs, but there are no cilia on the rest of the body. The alimentary canal is provided with a peculiar gizzard. The sexes are distinct, the males being frequently smaller than the females, and having the alimentary canal abortive.

The ECHINODERMATA may be divided into seven orders—

1. The ECHINIDEA (or Sea-urchins).—The body is spheroidal or discoidal. There are five main ambulacral tubes, which extend from the mouth to the apex, or summit of the opposite side of the body. The *corona* or main part of the skeleton is

composed of, 1. Series of plates which lie superficial to the ambulacral canals, and are called ambulacral, being perforated for the canals of the feet. 2. Plates which lie between these series, and are termed interambulacral. 3. Five ocular and five genital plates at the apical ends of the ambulacral and interambulacral series respectively.

The larvæ are pluteiform, and have a skeleton.

2. The HOLOTHURIDEA (or Sea-slugs).—The body is vermiform. The ambulacral tubes extend, as in the Echinidea, from mouth to apex, but there are no regular ambulacral and interambulacral plates. The genital organs have a single aperture towards the anterior part of the body.

The larvæ are vermiform, and have no skeleton.

3. The ASTERIDEA (or Star-fishes).—The body is pentagonal, or stellate, and depressed, the mouth being in the middle of one of the flat sides. The ambulacral tubes extend only to the extreme boundaries of the oral face of the discoidal body. Hard plates, or ossicles, articulated together, constitute the skeleton, and a double row of these lies on the deep side of the ambulacral canals, which, on the superficial side, are covered only by the integument.

The larvæ are vermiform, and have no skeleton.

4. The OPHIURIDEA (or Sand-stars).—The body is depressed, and gives off five arms of a different structure from itself. Each presents a central axis, formed by a chain of quadrate ossicles, and, for each ossicle, a zone of four superficial plates, one ventral, one dorsal, and two lateral. The ambulacral vessel lies between the ventral plates and the quadrate ossicles, and only extends to the summit of each arm.

The larvæ are pluteiform, and have a skeleton.

5. The CRINOIDEA (or Feather-stars).—The body is cup-like, sometimes stalked and sometimes sessile. It gives off a variable number of arms (usually five). The mouth is placed in the middle of the cup, and the principal plates of the skeleton, which are regularly articulated together, are developed on the opposite side of the body to that on which the mouth is situated. The arms are provided with articulated lateral processes or pinnules, clothed in a broad membrane, in which the reproductive organs are placed.

The embryo is free and ciliated, and developes within itself a second larval form, which becomes fixed by a peduncle.

The Crinoids are so different from the other living *Echinodermata* that they will probably have to form a distinct primary division, or sub-class of the class; and this may possibly be the case with some of the following extinct forms.

6. The CYSTIDEA.—The body spheroidal or ovoidal, inclosed in suturally united plates, and stalked. At the opposite end to the stalk is a terminal aperture, surrounded by arms; and on the side of the body another larger aperture, closed by a pyramid formed of triangular plates.

7. The EDRIASTERIDA.—The body depressed, hemispheroidal, inclosed in suturally united plates, but not stalked. In the centre of the convex face is an aperture from whence five ambulacra radiate: between two of these is a pyramid like that of the *Cystidea*.

8. The BLASTOIDEA.—The body is inclosed within suturally united plates, and stalked. There is no pyramid. A central aperture appears opposite the stalk, whence five ambulacra with a very complicated structure radiate; in four of the re-entering angles, between the ambulacra, is a pair of small apertures; and in the fifth, two small apertures, with a larger one between them, are placed.

These three groups have been extinct since the latter part of the palæozoic epoch.

GLOSSARY.

ABDOMEN (*abdere*, to conceal), in *Mammalia* the part of the trunk below the diaphragm; in *Insecta* the third coalesced segment of the body of the *Imago*.

ACALEPHÆ (ἀκαλήφη, a nettle), jelly-fish: they formed a class of Cuvier's Radiata, and are now included under *Hydrozoa*.

ACANTHOCEPHALA (ἄκανθα, a thorn; κεφαλή, head), a group of Scolecida.

ACARINA (ἄκαρι, a mite—Arist. Hist. An. v., xxxii. 2), an order of Arachnida.

ACETABULUM, a vinegar bowl (*acetum*), applied to the socket of the hip-joint in *Vertebrata*, and to the suckers on the tentacles of *Cephalopoda*.

ACTINOZOA (ἀκτίς, a ray; ζῶον, an animal), a class of Cœlenterata. Cf. *Anthozoa*.

ALISPHENOID (*ala*, a wing; σφήν, wedge; εἶδος, form; the lateral part of the second cranial segment, the greater wings of the sphenoid bone in Man.

ALLANTOIS (ἀλλᾶς, a sausage), a fœtal membrane of Amniotic Vertebrata.

ALVEOLUS (dim. of *alvus*, belly), a hollow cavity, specially applied to the sockets of teeth.

AMBULACRUM, a garden walk (*ambulare*), applied to the perforated spaces which run from the apex to the base of a sea-urchin.

AMŒBA (ἀμοιβή, a change), one of the Protozoa which is continually changing its form: hence it was formerly called Proteus.

AMPHIBIA (ἀμφί, both; βίος, life), living on land and water, and so applied by Linnæus to Reptiles and by others to Seals (*Pinnipedia*), but now recognised as the title of a class of Vertebrata, Cuvier's *Batrachia*, which at some time or other can breathe either in air or water.

AMPHIOXUS (ἀμφί, both; ὀξύς, sharp), the generic name of a fish, the only member of the order Pharyngobranchii, which has no enlargement of the brain or skull, and therefore tapers nearly equally at each end.

AMPHIPODA (ἀμφί, both; πούς, foot), a group of *Crustacea* in which the feet served both for walking and swimming, included under *Edriophthalmia*.

ANALOGY (ἀναλογία, proportion), the relation between parts which agree in function; as the wing of a bird and a butterfly, the tail of a whale, a fish and a sagitta.

ANKYLOSIS (ἀγκύλωσις)—sometimes written anchylosis—in anatomy, the coalescence of two bones so as to prevent motion between them.

ANNELIDA or "Annulata" (*annellus*, dim. of *anulus* or *annulus*, a ring), a class of *Annulosa*.

ANNULOSA (*annulus*, a ring), a primary division of the sub-kingdom, corresponding with Cuvier's *Articulata*.

ANOPLOTHERIUM (ἀ privative; ὅπλον, a weapon), an extinct genus of mammals unprovided with horns.

ANTENNA (the yard of a ship), applied to the appendages of one somite of the head in Insects, and two in Crustacea, the anterior in the latter case being distinguished as *Antennules*.

ANTHOZOA (ἄνθος, a flower; ζῶον, an animal), a class of Cœlenterata, otherwise called *Actinozoa*, and formerly *Zoophyta*, or animal plants, corresponding partly to Cuvier's *Polypi*.

APTERYX (ἀ priv.; πτέρυξ, a wing), the wingless bird of New Zealand, one of the *Ratitæ*.

ARACHNIDA (ἀράχνη, a spider), a class of arthropodous *Annulosa*.

ARANEINA (*aranea*, a spider), an order of *Arachnida*.

ARCHÆOPTERYX (ἀρχαῖος, ancient; πτέρυξ, a wing), an extinct genus of birds, the only representative of the order *Saururæ*.

ARCTISCA (ἄρκτος, a bear), an order of *Arachnida*, known as water-bears, otherwise called *Tardigrada*.

ARTHROGASTRA (ἄρθρον, a joint; γαστήρ, belly), an order of *Arachnida* with distinctly jointed abdomens.

ARTHROPODA (ἄρθρον, a joint; πούς, a foot), those Annulose classes which have jointed limbs.

ARTICULATA (*articulus*, dim. of *artus*, a joint), animals with jointed bodies, otherwise called *Annulosa*.

ARTIODACTYLA (ἄρτιος, even; δάκτυλος, a finger or toe), an order of *Mammalia*, with the toes of the feet either two or four in number: it includes the Ruminantia and some of the Pachydermata of Cuvier.

ASCARIS (ἀσκαρίς, a maw-worm), the "round-worm," one of the human Entozoa.

ASCIDIOIDA (ἀσκός, a bag), an order of Molluscoida, otherwise called *Tunicata*.

ASTERIDEA (ἀστήρ, a star, a star-fish—Arist. H. A. v., xv. 20), an order of *Echinodermata*.

ASTRAGALUS (ἀστράγαλος, a huckle-bone), the tarsal bone which articulates with the tibia.

ATLAS ("Ατλας, the god who holds up the earth), the vertebra which articulates with the skull.

ATRIUM (a hall), applied to the cavities peculiar to the *Molluscoida*.

AURICLE (*auricula*, dim. of *auris*, an ear), applied to an appendage of the heart in Mammals, and thence extended to the cavity of a heart of whatever form which receives blood and propels it into the ventricle. (This cavity is generally called *atrium* in German works.)

AXILLA (the long form of *ala*, a wing, as *maxilla* of *mala*), the arm-pit.

AXIS (ἄξων), a pivot: the second cervical vertebra on which the skull and atlas commonly rotate.

AZYGOS (ἀ priv.; ζυγόν, a yoke, a pair), an organ in the middle line of a bilaterally symmetrical animal, which has therefore no fellow.

BALEEN (*balæna*, φαλλαίνα, a whale—Arist. H. A. i., v. 2), the horny plates which line the roof of the true whales' mouths, commonly called whalebone.

BASIOCCIPITAL (*basis*, *ob*, *caput*), the centre of the first cranial segment, corresponding with the basilar process in human anatomy.

BASISPHENOID (βάσις, σφηνοειδής, wedge-shaped), the centre of the second cranial segment of the skull, corresponding with the body of the sphenoid behind the *Sella Turcica* in Man.

BATRACHIA (βάτραχος, a frog), an order of Cuvier's system, which corresponds with the class *Amphibia*; now used as an order of this class, corresponding with *Anura*, or tail-less Amphibia: *v.* URODELA.

BELODON (βέλος, a dart; ὀδούς, a tooth), an extinct genus of *Crocodilia*.

BLASTODERM (βλαστός, a sprout; δέρμα, skin), the superficial surface of the embryo in its earliest condition.

BLASTOIDEA (βλαστός, a sprout; εἶδος, appearance), an extinct order of *Echinodermata*.

BOTHRENCEPHALUS (βόθρος, a pit; κεφαλή, head), the broad-worm, a genus of *Tæniada*.

BRACHIOPODA (βραχίων, the arm; πούς, foot), a class of *Molluscoida*, formerly called *Palliobranchiata* from a mistaken theory that their mantle acts as a respiratory organ.

BRACT (*bractea*, a thin layer of metal or wood), applied in zoology to part of a hydrozoon, from its resemblance to the bract or sheath of flowers.

BRADYPUS (βραδύς, slow; πούς, foot), a genus of *Edentata*, the sloths.

BRANCHIOGASTEROPODA (βράγχια, gills—Arist. H. A. viii, ii. 8; γαστήρ, belly; πούς, foot), a class of gasteropodous mollusks which breathe by gills.

BRANCHIOPODA (βράγχια, gills; πούς, foot), an order of *Crustacea* in which the feet are modified to serve for respiration. It includes the orders *Phyllopoda* and *Cladocera* of Milne-Edwards.

BRANCHIO-STEGAL rays (βράγχια, gills; στέγω, to cover), parts of the hyoid apparatus in fish which support the membrane covering the gills.

BYSSUS (βύσσος, fine linen), silky threads secreted from the mantle of some Mollusks.

CADUCOUS (*cado*, fall), formed so as to fall off.

CÆCUM (sc. *intestinum*), a blind process of the alimentary canal.

CALCANEUM, or *os calcis* (*calx*, the heel), the bone of the *tarsus* which forms the prominence of the heel or hock.

CANINE (*dens caninus*), the eye-tooth, well developed in dogs and other carnivora, defined as that which comes next to the præmaxillary suture in the upper jaw of Mammals and that of the mandible which shuts in front of it.

CARINATÆ (*carina*, a keel), an order of Birds which have the sterna raised into a median ridge or keel.

CARNIVORA (*caro*, flesh; *vorare*, to devour), an order of Mammalia.

CARPUS (καρπός, the wrist), the bones between the forearm and the hand, usually eight in number in the higher *Vertebrata*; the wrist in Man, the "knee" in the Horse.

CELL (*cella*, a small room), often applied to any small cavity, but properly restricted to a microscopic anatomical element, with a nucleus, cell-wall and cell-contents, when typically formed.

CENTRUM (κέντρον, a prick, made by one leg of a pair of compasses), a centre, applied specially to the "bodies" of vertebræ.

CEPHALOPODA (κεφαλή, head; πούς, foot), a class of Mollusks.

CEPHALOTHORAX (κεφαλή, head; θώραξ, breast), the united head and thorax of *Crustacea* and *Arachnida*.

CERCARIFORM, like the Cercaria (κέρκος, tail), an aquatic worm which is one stage in the development of a Trematode fluke.

CETACEA (κῆτος, a whale), an order of Mammalia.

CHÆTOGNATHA (χαίτη, hair; γνάθος, the jaw), a class of Annulosa of which the genus *Sagitta* is the only representative.

CHÆTOPHORA (χαίτη, hair; φέρω, bear), an order of Annelida provided with bristles.

CHEIROPTERA (χείρ, hand; πτερόν, wing), an order of *Mammalia*.

CHELÆ (χηλή, claw), the modified fourth pair of thoracic limbs in Lobsters and their allies, the modified mandibles in Scorpions.

GLOSSARY.

Cheliceræ (χηλή, claw; κέρας, horn, antenna), the modified antennæ of Scorpions.
Chelonia (χελώνη, a tortoise), an order of Reptiles.
Chiasma (χίασμα, a crossing, fr. χιάζω, to mark with a χ), the commissure of the optic nerves which takes place in most *Vertebrata*.
Chilognatha (χεῖλος, lip; γνάθος, jaw), an order of Myriapoda, otherwise known as *Diplopoda*.
Chilopoda (χεῖλος, lip; πούς, foot), an order of Myriapoda, otherwise *Syngnatha*.
Chimæra (χίμαιρα, a monster like a goat), a genus of Elasmobranchiate fishes which has been made into a separate family, the *Holocephali*.
Chitin (χιτών, a tunic), the horny covering of Insects and other *Annulosa*.
Chlorophyll (χλωρός, green; φύλλον, leaf), the colouring matter of leaves.
Chorion (χόριον, a skin—Lat. *corium*), the vascular membrane which surrounds a vertebrate fœtus.
Cicatricula (dim. of *cicatrix*, a scar), the opaque spot on the surface of a fecundated yolk.
Cilia (pl. of *cilium*, an eyelash), microscopic filaments which move rhythmically.
Cirripedia (*cirrus*, a curl of hair; *pes*, a foot), a group of *Crustacea* in which several of the legs become cirrous: placed by Cuvier among the *Mollusca*.
Clavicle (*clavicula*, dim. of *clavis*, a key), the collar-bone of the pectoral arch or shoulder girdle in *Vertebrata*, in birds called the merrythought.
Cloaca (a sewer), the common duct into which the rectum, urethra and genital canals open in the *Sauropsida* and *Monotremata*.
Cochlea (κοχλίας, a snail), applied to part of the internal ear in the higher *Vertebrata*, from its spiral shape in Man.
Cœlenterata (κοῖλος, hollow; ἔντερα, viscera), a group of animals which differ from those below them in having a hollow digestive cavity.
Cœnosarc (κοινός, common; σάρξ, flesh), the common stem of a hydroid polypidom.
Cœnurus (κοινός, οὐρά, tail), the hydatid form of the wandered scolex of the dog's tapeworm with its deuto-scolices attached.
Coleoptera (κολεός, a sheath; πτερόν, a wing), an order of Insects with the membranous wings (second pair) enclosed in sheaths formed by the *elytra*.
Columella (dim. of *columna*, a pillar), the bone of the ear present in several Amphibia and most Sauropsida, which answers to the *stapes* in *Mammalia*.
Condyle (κόνδυλος, a knuckle), the articular surface of a bone, especially of the *occiput*.
Copepoda (κώπη, an oar; πούς, foot), an order of *Crustacea*, including, beside the group so named by Latreille and Milne-Edwards, most of the *Epizoa* or *Ichthyophthira*.
Coracoid (κόραξ, a crow, from its resemblance to a crow's beak in man), the second clavicle found in Reptiles, Birds, and Monotremata.
Coralligena (κοράλλιον, coral; γεν-, root of γίγνομαι, produce), an order of *Actinozoa*.
Corpus Callosum, "the firm body," the great transverse commissure of the cerebral hemispheres in *Mammalia*.
Corrodentia (*con* and *rodere*, gnaw), the family of insects to which the destructive "white ants" belong.
Cortical (*cortex*, bark), external—opposed to medullary.
Cotyledon (κοτυληδών), a cup-shaped hollow, applied in anatomy to the tufts of a ruminant placenta.
Craspedote (κρασπεδόω, to surround with a border), "the naked-eyed" *Medusæ*.

GLOSSARY.

CRINOIDEA (κρίνον, a lily), an order of *Echinodermata*, also called Pinnigrada.

CRUSTACEA (*crusta*, a shell), a class of *Annulosa*.

CTENOPHORA (κτείς, a comb; φέρω, to bear), an order of *Actinozoa*.

CUTIS (skin), applied to the vascular layer, or true skin, which is also called *corium* and *derma*, as distinguished from the scarf-skin, cuticle, or *epidermis*.

CYCLOBRANCHIATA (κύκλος, a ring; βράγχια, gills), a family of *Gasteropoda*.

CYSTICERCUS (κύστις, bladder; κέρκος, tail), the wandered scolex of *Tænia solium*, in its hydatid form.

DECIDUA (sc. *membrana*, *de* and *cado*), the modified mucous membrane of the pregnant uterus, when it falls off at birth.

DEMODEX (δημός, fat; δήξ, a worm), a minute Arachnid allied to the *Pentastomida*, and inhabiting the sebaceous follicles of Man.

DERMATOPTERA (δέρμα, skin; πτερόν, wing), a family of insects with membranous wings.

DIAPHRAGM (διάφραγμα, otherwise called φρήν, whence the adj. phrenic), the muscle which in *Mammalia* separates the abdomen from the thorax.

DIASTEMA (διά, apart; ἵστημι, to place), an interval, especially between teeth.

DIATOMACEA (διατέμνω, to cut through), the siliceous coverings of a large group of microscopic low vegetable organisms.

DICYNODON (δι-, two; κύων, a dog; ὀδούς, tooth), an extinct reptile with two canine-like teeth.

DIDELPHIA (δι-, two; δελφύς, the womb), the subclass of Marsupial Mammals.

DIGITIGRADA (*digitus*, a finger; *gradere*, to walk), a term applied to those mammals which walk on the phalanges of their fore and hind feet.

DINOSAURIA (δεινός, terrible; σαύρα, lizard), an extinct order of reptiles.

DIŒCIOUS (δι-, two; οἶκος, a house), with distinct sexes.

DIPNOI (δι-, double; πνοή, breath), an order of Fishes which breathe both by lungs and gills, otherwise known as *Diplopnoi* or *Protopteri*.

DISCOPHORA (δίσκος, a quoit or disk; φέρω, bear), an order of *Annelida*, also called *Suctoria*, corresponding with the family of Leeches (*Hirudinea*).

ECDYSIS (ἔκδυσις, fr. ἐκδύω, to put off), casting the skin, or moulting.

ECHINOCOCCUS (ἐχῖνος, an urchin; κόκκος, a berry), the wandered scolex of *Tænia echinococcus*, in its hydatid form, with deuto-scolices, or daughter cysts formed by gemmation.

ECHINODERMATA (ἐχῖνος, a hedgehog or urchin, hence applied to the Sea-urchin, and δέρμα, skin), a class represented by the Sea-urchin.

ECHINORHYNCHUS (ἐχῖνος, a hedgehog; ῥύγχος, snout), a genus of *Acanthocephala*.

EDENTATA (*e* and *dens*), without teeth, an order of Mammalia in which the teeth are wholly or partially absent.

EDRIOPHTHALMIA, or *Hedræophthalmia* (ἑδραῖος, sessile; ὀφθαλμός, eye), an order of *Crustacea*, including the *Isopoda*, *Amphipoda*, and *Læmodipoda*.

ELASMOBRANCHII (ἔλασμα, a thin plate or lamina; βράγχια, gills), an order of Fishes, including the *Desmiobranchii Plagiostoma* or *Pentabranchii* of Cuvier, or *Placoidei* of Agassiz), and *Holocephali*.

ELYTRUM (ἔλυτρον, fr. ἐλύω, to roll round—Arist. H. A. iv., vii. 8), the hard first pair of wings which in Beetles cover the second pair.

EMBRYO (ἔμβρυον, fr. ἐν and βρύον, fr. βρύω, swell), the earliest stage in which an animal appears in impregnated ovum: later, especially among Vertebrata, it is called *fœtus*.

ENDO-PODITE (ἔνδον, within; πούς, foot), the internal distal segment of the typical limb of *Crustacea*.

ENDOSTYLE (ἔνδον, within; στῦλος, a pillar), a fold of the lining membrane of the pharynx in *Ascidioida*.

ENTOMOSTRACA (ἐντέμνω, to cut up; ὄστρακον, a shell, *crusta*), the lower *Crustacea*, so called by Latreille from the segments of their bodies being unconsolidated.

EPHEMERIDÆ (ἐφήμερον, fr. ἐπί and ἡμέρα, an insect that lives a single day, Arist. H. A. v. 12, 26), a family of Insects.

EPIOTIC (ἐπί, upon; οὖς, ear), the upper bone of the auditory capsule, part of the *pars petrosa* in Man.

EPIPODITE (ἐπί, upon; πούς, foot), the external distal segment of the typical limb of *Crustacea*.

ERRANTIA (*errare*, to wander), a family of Annelida fitted for locomotion, included under *Chætophora*.

EX-OCCIPITAL (*ex*, and *occiput*), the lateral parts of the first cranial segment, corresponding with the sides of the *Foramen magnum* in Man.

FEMUR, the thigh-bone, between the *pelvis* and *tibia* of the posterior limb of the *Amphibia*, *Sauropsida*, and *Mammalia*.

FIBULA (a brooch, περόνη), the smaller external bone of the leg in the higher *Vertebrata*, homologous with the *ulna* in the fore-arm.

FIMBRIA, a fringe.

FISSIPAROUS (*fissus*, cleft; *pario*, to bring forth), asexual generation by the parent splitting into two parts, which become new individuals.

FLAGELLUM (a whip), the whip-like appendage to the *Pilidium*, q. v.

FORAMINIFERA (*foramen*, a hole; *fero*, to bear), a group of *Rhizopoda* which live in hollow, perforated, calcareous shells: to it belong the *Orthocerina*, *Globigerina*, and *Nummulites*.

FRONTAL (*frons*, forehead), the upper part of the third cranial segment, corresponding to the vertical part of the frontal bones in Man.

GALEOPITHECUS (γαλέη, a weasel; πίθηκος, an ape), a genus of Cologus, or flying cats, placed by Linnæus among the Lemurs, by Cuvier among the Bats, and by Geoffroi St. Hilaire in the group *Carnassiers*, to the Insectivorous order of which it properly belongs, forming a family which has been named *Dermoptera*.

GANGLION (γάγγλιον, a swelling or lump), in anatomy a centre of the nervous system, containing nerve cells, and receiving and giving out impressions.

GANOIDEI (γάνος, brightness), an order of fishes mostly extinct, the *Sauroidei* of Agassiz, *Holostei* of Müller.

GEMMIPAROUS (*gemma*, bud; *pario*, bring forth), asexual generation by new individuals arising as buds from the body of the parent.

GEMMULE (*gemmula*, dim. of *gemma*, a bud), an encysted mass of sponge-particles, from which new ones are produced.

GEPHYREA (γέφυρα, a bridge), a group which includes the families *Sipunculidæ* and *Synaptidæ*, placed as an order of *Echinodermata* called *Apedicellata* (Cuvier), *Apoda* (Van der Hoeven), or *Vermigrada* (Forbes), but now admitted among the *Annelida*.

GLAND (*glans*, an acorn), originally any smooth round viscus, now properly restricted to those viscera which "secrete," *i. e.*, separate certain constituents of the blood by a process of cell-growth, which are afterwards poured out from the gland by a duct.

GREGARINA (*gregarius*, fr. *grex*, a flock, occurring in numbers together), a genus of Protozoa which forms the type of the class *Gregarinida*.

GORDIUS, a genus of Nematoid worms, giving its name to the group *Gordiacea*, and named, from its wreathing its thread-like body into knots, after the mythical king of Phrygia the knot of whose waggon was cut by Alexander.

GYMNOLÆMATA (γυμνός, naked; λαιμός, throat), an order of Polyzoa in which the opening to the gullet is uncovered.

GYMNOPHIONA (γυμνός, naked; ὄφις, snake), an order of Amphibia, otherwise called *Ophiomorpha* or *Cæciliæ*.

GYMNOPHTHALMATA (γυμνός, naked; ὀφθαλμός, eye), the craspedote *Medusæ* of the class *Hydrozoa*.

HALLUX (*hallex* or *allex*, the great toe), the innermost of the five normal digits of a vertebrate foot.

HALTERES (ἀλτῆρες, fr. ἅλλομαι, to leap), poisers, weights held in the hand in leaping, applied to the modified second pair of wings of *Diptera*.

HELMINTHLÆ (ἕλμινς, a worm, fr. εἴλω, to twist), a synonym of Entozoa, divided by Owen into *Sterelmintha* (στερεός, solid), the *Parenchymata* of Cuvier, and *Cœlelminthia* (κοῖλος, hollow), the *Cavitaria* of Cuvier. They are all included in the class *Scolecida*.

HEMIPTERA (ἥμι-, half; πτερόν, a wing), an order of Insects with the anterior wings half coriaceous.

HETEROPODA (ἕτερος, different from others; πούς, foot), a group of branchial Gasteropoda in which the *propodium* is turned into a laterally compressed fin, while the *epipodia* are absent.

HOLOTHURIA (ὁλοθούριον—Arist. H. A. i., i. 19), a genus to which the Sea-cucumbers and Trepangs belong, which gives its name to the order *Holothuridea*.

HOMOLOGY (ὁμολογία, agreement), the relation between parts which are developed out of the same embryonic structures; as the arm of a man, the foreleg of a horse, and the wing of a bird, or the ' wings " of a pteropod, and the tentacles of a cuttle-fish. The term *Serial Homology* is applied to the likeness between parts which appear to be the modified development of structures similarly repeated, as the humerus and femur in Vertebrata, or the maxillæ, maxillipeds, and ambulatory limbs of Crustacea.

HUMERUS (*brachium*, βραχίων), the bone of the upper arm in *Vertebrata*.

HYDATID (ὑδατίς), or Bladder-worm, the cystic form of the wandered scolices of tape-worms.

HYDRA (ὕδρα, a water-dragon), a genus of Polyps first described by Trembley in 1774, which forms the type of the modern class *Hydrozoa*.

HYMENOPTERA (ὑμήν, a membrane; πτερόν, wing), an order of Insects with two pair of wings, both membranous.

HYOID (Y, εἶδος, resemblance), also called *os linguæ* from its supporting the tongue, a bone named from its resemblance to the letter U in Man: in most other animals it is much more complicated, the lesser cornua forming with the stylohyoid ligaments long jointed bones which connect it with the skull.

HYRAX (ὕραξ, *sorex*, a shrew), the Daman, the Coney (*i.e.*, Rabbit) of Scripture, a small gregarious mammal found in Syria and South Africa. Linnæus put it among *Rodentia*, Cuvier under *Pachydermata;* but its peculiarities are so great that it may form the type of a distinct order, *Hyracoidea*.

ICHTHYOPSIDA (ἰχθύς, a fish; ὄψις, appearance), a primary division of *Vertebrata*, which includes the classes *Pisces* and *Amphibia*.

ICHTHYOSAURUS (ἰχθύς, a fish; σαύρα, a lizard), an extinct genus of Reptiles, giving its name to the order *Ichthyosauria*.

IMAGO, applied by Linnæus to the final, winged and sexual state of Insects.

INCISOR (*incido*, to cut), the teeth (usually broad and sharp) fixed in the *præmaxilla* of Mammalia, and those in the lower jaw which shut against them.

INCUS (an anvil, fr. *incudo*), the bone of the ear in Mammalia which, according to Reichert, is the homologue of the *os quadratum* in lower vertebrates, but has been now shown to be represented by a ligament only in birds, a cartilage in *Sphenodon* (Hatteria) and other reptiles, and the hyo-mandibular bone in osseous fishes : *v.* MALLEUS.

INFUNDIBULUM (a funnel), the channel formed by folded processes of the mantle by which water passes out from the branchial chamber of Mollusks.

INFUSORIA (*infusum*, fr. *in* and *fundo*, pour), a class of microscopic animals, named from their occurrence in vegetable infusions.

INGUEN, the groin.

INSECTA (*inseco*, to cut in pieces, as in Greek, ἔντομα—Arist. II. A. iv. 1, 5—fr. ἐντέμνω), a class of Arthropodous *Annulosa*. "Jure omnia Insecta appellata ab incisuris, quæ nunc cervicum loco, nunc pecterum atque alvi, præcincta separant membra."—Plin. xi. 1, 1.

INSECTIVORA (*insectum*, an insect ; *voro*, eat), an order of Mammalia.

INTERMAXILLARY or præ-maxillary bones, so called because they are placed in front of and between the *maxillæ* of *Vertebrata*, with which they unite in adult life in Man and some of the apes.

ISOPODA (ἴσος, equal ; πούς, foot), an order of *Crustacea* included under *Edriophthalmia*.

JUGAL (*jugum*, a yoke), a bone of the face corresponding to the *os malæ* or cheek-bone in human anatomy, and forming part of the *zygoma*.

LABIUM (a lip), the lower part of the mouth in insects formed of the coalesced second part of maxillæ.

LABRUM (form of *labium*), the upper part of the mouth in insects.

LÆMODIPODA (λαιμός, throat ; δι-, two ; πούς, foot), a group of *Crustacea* with two legs under the throat, now included under *Edriophthalmia*.

LAMELLIBRANCHIATA (*lamella*, dim. of *lamina*, a thin plate ; βράγχια, gills), an order of *Mollusca*.

LARVA (a mask), the name applied by Linnæus to the grub or caterpillar, which is the first stage in the metamorphosis of Insects.

LEPIDOPTERA (λεπίς, scale ; πτερόν, wing), an order of Insects with wings covered with scales.

LEPIDOSIREN (λεπίς, scale ; σειρήν, a siren—applied to one of the *Amphibia*), a fish resembling the Siren, but covered with scales.

LEPIDOSTEUS (λεπίς, scale ; ὀστέον, bone), the Bony Pike of North America.

LOPHOPHORE (λόφος, a plume ; φέρω, bear), the oral disk of *Polyzoa*.

LUCERNARIA (*lucerna*, a lamp), a genus of *Hydrozoa*.

MACRAUCHENIA (μακρός, long ; αὐχήν, neck), an extinct genus of *Mammalia*.

MALACOSTRACA (μαλακόστρακα—Arist., soft-shelled, *v.* ENTOMOSTRACA), the higher Crustacea, so called by Latreille in distinction from the harder-shelled Mollusca.

MALLEUS (a hammer), a bone of the internal ear in *Mammalia*, which Reichert supposed to answer to the *pars articularis* of the mandible in Sauropsida, but which is probably the true homologue of the *os quadratum* : *v.* INCUS.

MALLOPHAGA (μαλλός, wool ; φαγεῖν, to eat), a group of parasitic Insects.

MAMMALIA (*mamma*, a breast), a class of *Vertebrata* distinguished by suckling their young.

MANDIBLE (*mandibulum*, fr. *mando*, to chew), in *Vertebrata*, the lower jaw, answering to the *maxilla inferior* of human anatomy : in *Arthropoda*, the upper pair of cephalic appendages used as jaws : among Birds, the upper and lower *rostra* of the beak are often called mandibles.

MARSIPOBRANCHII (μάρσιπος, a pouch ; βράγχια, gills), an order of Fishes, called *Cyclostomi* by Müller, represented by the Lampreys and Hags.

MARSUPIALIA (*marsupium*, a pouch), an order of *Mammalia* separated by Cuvier from the rest of his *Carnassiers*, and now forming the subclass *Didelphia*.

MASTODON (μαστός, breast ; ὀδούς, tooth), an extinct genus of *Proboscidea*, representing in America the Mammoths of the Old World, and distinguished by the nipple-like surface of their molar teeth.

MAXILLA (the long form of *mala*, a jaw—*v.* AXILLA), in *Vertebrata*, the bone corresponding with the Superior Maxilla of human anatomy : in *Arthropoda*, the one or two pair of limbs next to the mandibles which are modified as jaws.

MAXILLIPED (*maxilla, pes*), or "foot-jaw," applied to the modified limbs of the three first segments of the thorax in *Crustacea*.

MEDULLA (marrow), applied to the spinal part of the central nervous system in Vertebrata—the spinal cord.

MEGATHERIUM (μέγα, great ; θηρίον, beast), an extinct animal allied to the sloths.

MENISCUS (μηνίσκος, dim. of μήνη, a crescent), applied to an organ of doubtful function in *Echinorhynchus*.

MENTUM (chin), the central and lower part of the *labium* of insects.

MEROSTOMATA (μηρός, thigh ; στόμα, mouth), an order of *Crustacea* represented by the genus *Limulus*, the Molucca King-Crabs, alone in recent times : the order has been also called *Xiphosura*, or sword-tailed.

MESENTERY (μέσος, intermediate ; ἔντερον, intestine), the membrane between the alimentary canal and the walls of the abdomen. So Meso-colon, &c.

MESERAIC = mesenteric. The omphalo-meseraic vessels pass from the intestine to the umbilical vesicle in the embryo.

METACARPUS (μετά, after ; καρπός, wrist), the bones between the carpus and phalanges of the anterior limb of the higher Vertebrates.

METATARSUS (μετά, ταρσός, the flat of the foot), the bones between the tarsus and phalanges of the hind limb of the higher Vertebrates.

MOLAR (*molaris*, adj. of *mola*, a mill), applied to those teeth in Mammals which are not preceded by a milk set—the "grinders" in Man.

MOLLUSCA (fr. *mollis*, soft), one of the primary divisions of the Animal Kingdom established by Cuvier. He included in it, beside the classes now admitted, the *Tunicata, Brachiopoda*, and *Cirripedia*.

MONODELPHIA (μόνος, single ; δελφύς, womb), the orders of *Mammalia* with a single *uterus*.

MONŒCIOUS (μόνος, single ; οἶκος, a house), with both sexes united in one individual, *hermaphrodite*.

MONOTREMATA (μόνος, single ; τρῆμα, fr. τιτραίνω, to pierce), an order of Mammals, separated from Cuvier's *Edentata*, and forming the subclass Ornithodelphia : they have only one opening for the urinary, genital and intestinal canals.

MYRIAPODA (μυρίος, numerous ; πούς, foot), an order of Arthropodous *Annulosa* separated from the *Insecta* of Cuvier, and represented by the Millipedes and Centipedes.

NECTOCALYX (νηκτός, fr. νήχω, swim ; κάλυξ, cup), the swimming-bell of *Hydrozoa*.

NEMATOCYST (νῆμα, a thread ; κύστις, a bladder), the "thread-cells" found in all *Cœlenterata* and some un-allied genera of *Turbellaria* and *Mollusca*.

NEMATOIDEA (νῆμα, thread ; εἶδος, appearance), an order of *Scolecida* corresponding nearly with the *Cælelminthia* or *Entozoa Cavitaria* of Cuvier and the *Nematelminthiæ* of Vogt. It includes the parasitic "Thread Worms" and "Round Worms," and the *Gordiacea* may be placed under the same head.

NERVE (*nervus*, νεῦρον), originally a sinew or tendon: after Galen's time applied to the conducting branches of the Sensori-motor apparatus.

NEUROPTERA (νεῦρον, a cord; πτερόν, wing), an order of insects in which the four membranous wings are supported by strong ribs or nervures.

NOTOCHORD (νῶτον, back; χορδή, a string), or *Chorda dorsalis*, an embryonic structure in *Vertebrata* formed immediately under the primitive groove, and usually replaced by the spinal column in the adult.

NUCLEUS (a kernel, fr. *nux*: dim. *nucleolus*), a speck of germinal matter found normally in cells.

ODONTOIDES (sc. *processus*; ὀδούς, tooth; εἶδος, form), the body of the atlas, which is separate from it and usually ankylosed with the axis.

ODONTOPHORA (ὀδούς, tooth; φέρω, bear), those classes of *Mollusca* which have heads and a peculiar tooth-bearing apparatus.

ŒSOPHAGUS (οἶσος, a reed; φαγεῖν, to eat), the gullet—*Arist. de Part. An.* iii. 3.

OMPHALOS (ὀμφαλός=umbilicus), the navel, *i.e.*, the scar left in the abdomen of a mammal, where the umbilical cord was attached.

OOSTEGITE (ὠόν, egg; στέγω, to cover), scales or other parts of *Annulosa* modified so as to protect the eggs while carried by the mother.

OPERCULUM (fr. *operio*), a covering: in fish a bony flap (possibly homologous with the external ear of Mammals) which covers over the gills; in univalve Mollusks a concretion which closes the shells.

OPHIDIA (ὄφις, a serpent), an order of Reptiles: *Serpentes* of Linnæus.

OPHIURA (ὄφις, snake; οὐρά, a tail), "Brittle star," a genus of *Echinodermata*, giving its name to the order *Ophiuridea*.

OPISTHOTIC (ὄπισθεν, behind; οὖς, the ear), the posterior ossification of the auditory capsule, corresponding with the mastoid and part of the petrous bones in Man.

OPTIC LOBES, the ganglia of the brain in *Vertebrata* to which the optic nerves lead in all but the *Marsipobranchii* and *Amphioxus*: they are single on each side in the lower classes, and called *Corpora bigemina*; double in the higher ones, and called *C. quadrigemina*, or *Nates* and *Testes*.

ORBITOSPHENOID (*orbitus*, dim. of *orbs*; *sphenoides*), part of the third cranial segment, corresponding with the *alæ minores* or processes of Ingrassias in human anatomy, and always forming the back of the orbit. Cf. ALISPHENOID.

ORNITHODELPHIA (ὄρνις, bird; δελφύς, womb), the subclass represented by the order *Monotremata*.

ORNITHORHYNCHUS (ὄρνις, a bird; ῥύγχος, a beak), a genus of *Monotremata*, otherwise called the Duck-billed Platypus (broad, *i.e.*, webbed-foot), or Water Mole.

ORTHOPTERA (ὀρθός, straight; πτερόν, wing), an order of Insects.

OSTRACODA (ὀστρακώδης, adj. fr. ὄστρακον, a shell), an order of *Crustacea* enclosed in a hard carapace.

OTOLITHS (οὖς, ear; λίθος, stone), small bones found suspended in the internal ear of fishes, corresponding with the otoconium or "ear-dust" of Man. The term is also applied to similar concretions in the auditory sacs of *Crustacea* and other invertebrate animals.

OXYURIS (ὀξύς, sharp; οὐρά, tail), thread-worm, one of the Nematoidea.

PACHYDERMATA (παχύς, thick; δέρμα, skin), a Mammalian order of Cuvier's, nearly agreeing with the *Belluæ* of Linnæus. Of its members the Elephants now form the order *Proboscidia*, the Hyrax *Hyracoidea*, and the remainder may be called *Ungulata*, the artiodactylous genera uniting with the Ruminants.

PALÆOTHERIUM (παλαιός, ancient; θηρίον, beast), a Tapir-like ungulate mammal of the Tertiary epoch.

PALLIUM (a cloak), the "mantle" of Mollusks, an extreme development of the integument, represented in its epithelial, vascular, glandular, and muscular structure, with folds and processes forming the "foot" and other appendages.

PARASPHENOID (παρά, beside; σφηνοειδής, wedge-shaped), a long azygos bone which runs from before backward under the base of the skull in *Ichthyopsida* and some *Reptiles*: so called from its relation to the sphenoid bone.

PARENCHYMA (παρέγχυμα, fr. παρά, ἐν, χύω, something poured in beside), applied to the proper substance of viscera, excluding connective tissue, blood-vessels, and other accessory parts.

PARIETAL (*parietes*, walls), the upper ossifications in membrane of the second cranial segment.

PARIETO-SPLANCHNIC (*parietes*, σπλάγχνα, the viscera), a ganglion in the higher Mollusks which supplies the mantle, gills, and viscera.

PARTHENOGENESIS (παρθένος, virgin; γένεσις, generation), asexual reproduction, either by fission, gemmation, or the process of internal budding best seen in the Scolecida, and called by Quatrefages "Genea-genesis."

PATAGIUM (παταγεῖον), a stripe or border to a dress; applied to the expansion of the integument by which bats, flying squirrels, flying lemurs, &c., support themselves in the air.

PAUROPUS (παῦρος, few; πούς, foot), a genus of *Myriapoda* intermediate between the two orders represented by the *Scolopendridæ* and *Iulidæ*.

PECTOSTRACA (πηκτός, fr. πήγνυμι, fixed, compacted; ὄστρακον, a shell), an order of *Crustacea*, which become fixed in the adult state, including the *Cirripedia*, and *Rhizocephalus* with its allies.

PELVIS (πέλις, πέλυς), a basin, applied to the "hip-girdle" or bony arch supporting the lower extremities of most *Vertebrata*: it consists of the *sacrum* (formed of two to five vertebræ), the *ilia* or haunch bones, *pubes*, and *ischia*.

PENTASTOMA (πέντε, five; στόμα, mouth), a name given under a wrong conception to a genus of parasitic habit, otherwise called *Linguatula*, which gives its name to an order of *Arachnida*.

PERICARDIUM (περί, around; κάρδια, the heart), the membrane which surrounds the heart: in *Crustacea* this is really a venous sinus.

PERISSODACTYLA (περισσός, uneven; δάκτυλος, toe), a name applied to those Ungulate Mammals which have an odd number of digits, as the Tapir and the Horse.

PERITONEUM (περί, around; τείνω, to stretch), the membrane which covers the abdominal walls and the contained viscera.

PHALANX (φάλαγξ, a rank or row), one of the small bones which compose the digits of the higher *Vertebrata*, otherwise called *Internodii*.

PHARYNX (φάρυγξ), the upper part of the gullet.

PHARYNGOBRANCHII (φάρυγξ, βράγχια, gills), the order of Fishes represented by the *Amphioxus*, in which the perforated pharynx performs the function of gills: otherwise called *Leptocardii*.

PHYLACTOLÆMATA (φυλακτός, fr. φυλάσσω, guarded; λαιμός, throat), an order of *Polyzoa* in which the entrance to the gullet is guarded by an epistome.

PHYSOPHORIDÆ (φῦσα, bellows; φέρω, bear), a family of Hydrozoa.

PHYSOPODA (φῦσα, bellows; πούς, foot), a group of insects with bladder-like feet.

PILLIDIUM (for *pileolus*, dim. of *pileus*, a felt cap), the name applied by Müller to the helmet-shaped larva of Nemertes, one of the *Turbellaria*.

PINEAL body or "gland" (*pinna*, a fir-cone), otherwise called *Conarium*; a constant outgrowth from the roof of the Prosencephalon of *Vertebrata*.

PINNULE (*pinnula*, dim. of *pinna* or *penna*, a feather), the secondary branches from the quills of a feather, otherwise called barbs.

PITUITARY body or "gland" (*pituita*, phlegm, which it was supposed to secrete), a constant appendage of the brain just in front of the Notochord.

PLACENTA (a cake), applied to the developed *chorion* in Mammalia from its discoid shape in Man.

PLACOIDEI (πλάξ, a flat plate; εἶδος, likeness), an order of fishes with flat smooth integument, the *Selachii*, or Sharks and Rays: *v.* ELASMOBRANCHII.

PLANTIGRADE (*planta*, the sole of the foot; *gradior*, to walk), applied to those animals which apply the whole of the foot, including the heel, to the ground, as Man, bears, and badgers.

PLESIOSAURUS (πλησίος, near; σαύρα, a lizard), an extinct genus of marine reptiles which gives its name to an order, *Plesiosauria*, so called because their skeletons were first found associated with those of the *Ichthyosaurus*.

PLEURODONT (πλευρά, a rib, side; ὀδούς, tooth), the attachment of teeth to the jaw in which one side of the fang became ankylosed with the inside of the socket.

PLUTEUS (a pent-house, or shed), applied to the "painter's-easel" larva of the Echinus.

PODOPHTHALMIA (πούς, foot; ὀφθαλμός, eye), an order of *Crustacea* in which the eyes are stalked or pedunculate; it corresponds with the group *Decapoda*.

POLLEX, the thumb: the first, innermost, or most præaxial of the digits of the anterior extremity, placed in a line with the *radius*.

POLYCYSTINA (πολύς, many; κύστις, a bladder), the minute shells of *Radiolaria*.

POLYGASTRICA (πολύς, γαστήρ, belly), a name given by Ehrenberg under a wrong impression to *Infusoria*.

POLYZOA (πολύς, ζῶον, animal), a class of compound animals, otherwise called Bryozoa (βρύον, moss), from their parasitic habit on sea-weed.

PRÆAXIAL and POSTAXIAL are applied respectively to the parts on the radial or tibial and those on the ulnar or fibular side of the limbs; the former being internal or anterior, the latter external or posterior, to the axis of the limb.

PRÆMOLAR (*præ*, in front; *molares*, grinders), the molar teeth which are preceded by milk molars, the bicuspids.

PRÆSPHENOID (*præ*, before; *os sphenoidale*), the centrum of the third cranial segment, corresponding in human anatomy to the front part of the body of the sphenoid bone: *v.* BASISPHENOID.

PROGLOTTIS (προγλωττίς, the point of the tongue), applied to the zooids of *Scolecida* which are propagated by gemmation from a *scolex*, and in their turn produce *ova*,

PRO-OTIC (πρό, front; οὖς, ear), the anterior ossification of the auditory capsule, corresponding in human anatomy with part of the petrous bone.

PROPODITE (πρό, front; πούς, foot), the proximal segment of the typical limb of a Crustacean.

PROTOPLASM (πρῶτος, first; πλάσμα, fr. πλάσσω, to form), the primitive indifferent tissue of the embryo out of which all subsequent organs are formed by a process of differentiation.

PROTOZOA (πρῶτος, first; ζῶον, animal), the lowest group of animals. A similar term, " Protista," is applied by Häckel so as to include Protozoa and Protophyta.

PSEUDOPODIA (ψευδής, false; πούς, foot), the prolongations of the body thrust out and drawn in at will, which answer the purpose of limbs in *Rhizopoda*.

PTERODACTYLA (πτερόν, wing; δάκτυλος, finger), an order of extinct reptiles, characterised by the fifth digit of the anterior extremity being prolonged so as to support a patagium.

PTEROPODA (πτερόν, πούς, foot), a class of *Mollusca* in which the *Epipodia* of the foot are developed so as to form wing-like processes by which it swims.

PTERYGOID (πτέρυξ, wing; εἶδος, likeness), a bone of the vertebrate skull corresponding with the internal pterygoid processes in Man.

PULMOGASTEROPODA (*pulmo*, lung; γαστήρ, belly; πούς, foot), those *Mollusca* which walk on their bellies and breathe by lungs.

PUPA (a doll), applied to the second, usually motionless, stage of metamorphosis in Insects, otherwise called "Nymph" and Chrysalis.

PYCNOGONIDA or Pycnogonata (πυκνός, thick; γόνυ, knee); an order of *Arachnida* with thick jointed legs.

PYLORUS (πυλωρός, a gatekeeper), applied to the valve between the stomach and intestines.

PYRIFORM (*pyrum* or *pirum*, a pear; *forma*, shape), applied to any tapering organ.

QUADRATUM (sc. *os*), four-cornered, square; the bone by which the mandible articulates with the skull in *Sauropsida*. It probably answers to the Malleus in *Mammalia*.

RADIATA (*radius*, a spoke or ray), applied by Cuvier to a sub-kingdom now broken up. The *Polypi* are divided between *Polyzoa* and *Cœlenterata*, which last group takes also the *Acalephæ*; the *Entozoa* become *Scoleoida*, the *Echinodermata*, *Annuloida*, and only the *Infusoria* remain, minus the *Rotifera*.

RADIOLARIA (*radiolus*, dim. of *radius*), a class of *Protozoa*, of which the Sea-egg (*Sphærozoon ovo-di-mare*) is an example.

RADIUS (a spoke), the præxial bone of the fore-arm which articulates in a line with the pollex.

RAMUS (a branch), applied specially to each half of the mandible in *Mammalia*.

RATITÆ (*ratis*, a raft), an order of Birds (*Brevipennes* of Cuvier, *Cursores* of Illiger), so called because, other birds having a "keeled" sternum, their keelless one is like a punt or raft.

RHIZOCEPHALUS (ρίζα, a root; κεφαλή, a head), a genus of *Crustacea* which when adult bury their heads in the bodies on which they are parasitic.

RHIZODONT (ρίζα, ὀδούς, a tooth), the attachment of teeth whose fangs branch out, and so become ankylosed with the jaw-bone.

RHIZOPODA (ρίζα, πούς foot), a class of Protozoa in which pseudopodia come out of the body like roots.

RODENTIA (*rodere*, to gnaw), an order of Mammals, the *Glires* of Linnæus.

ROSTRUM (a beak), the anterior termination of the carapace in *Crustacea*.

ROTIFERA (*rota*, a wheel; *fero*, bear), or *Rotatoria*, a class of animalcules separated from the *Infusoria*, and provided with ciliated fringes round the mouth which when in motion look like two toothed wheels.

SACRUM (sc. *os*), the vertebræ which articulate with the *ilia* to form the pelvis.

SAGITTA (an arrow), a genus resembling some Annelida, but with peculiarities which have led to its being made into a separate class, "Chætognatha."

SARCODE (σάρξ, flesh; ὁδός, way), a name applied to the imperfectly differentiated tissue of Protozoa and Infusoria, which is as it were on its way to become true flesh.

SAUROPSIDA (σαύρα, a lizard; ὄψις, appearance), a name applied to the classes *Aves* and *Reptilia* collectively.

SAURURÆ (σαύρα, οὐρά, tail), an order of birds represented only by the extinct genus *Archæopteryx*, distinguished by having a long tail like a lizard's.

SCAPULA (form of *spatula*, dim. of *spatha*, σπάθη, a broad, flat blade), the shoulder-blade, called the "side-bone" in birds.

SCLEROTICA (σκληρός, hard, sc. *tunica*), the capsule of the retina. the eye-ball, fibrous in Man, but partially ossified in many of the lower *Vertebrata*.

SCOLECIDA, a group of *Annuloida* or *Vermes* comprehending the *Entozoa* of Cuvier, and also the free *Turbellaria*.

SCOLEX (σκώληξ, a worm), the larva in *Scolecida*, produced from an egg, which may by gemmation give rise to infertile *deutoscolices*, or to ovigerous *proglottides*.

SCUTUM (a shield), applied to the bony dermal plates in the skin of crocodiles, &c.: also the large dorsal scales of some *Annelida*.

SETIGEROUS (*setæ*, bristles; *gero*, carry), especially applied to the locomotive *Annelida*.

SIRENIA (σειρήν, a siren, a mermaid), an order of Mammals, the Herbivorous Cetacea of Cuvier, including the genera *Halicore* (dugong), *Manatus* (lamantin), and the recently extinct *Rhytina*; the name is given because the dugong, from its pectoral mammæ, was named Halicore, the sea-maiden.

SOMITE (σῶμα, body), a segment of the body of *Annulosa*, with its upper and lower pair of appendages.

SPERMATOZOA (σπέρμα, seed; ζῷον, animal), *animalcula seminis*, minute organisms of characteristic shape, and endowed with spontaneous motion, found in the sperm-cells of all animals.

SPICULUM (dim. of *spica*, a thorn), any hard, pointed animal structure.

SPIRACLE (*spiro*, to breathe), the lateral openings into the tracheal tubes of insects, &c.

SPONGIDA (σπογγιά or σπόγγος, a sponge—Arist. H. A. ix., xiv. 3), a class of *Protozoa*.

SQUAMOSAL (*squama*, a scale), a membrane bone, wedged in between the auditory capsule and the ali-sphenoid: in Man it overlaps the parietal by a scale-like suture.

STAPES (a stirrup), so called from its shape in Man; the auditory ossicle which is joined to the *Fenestra ovalis*, and corresponds with the Columella in *Sauropsida*.

STEGANOPHTHALMOUS (στεγανός, covered; ὀφθαλμός, eye), the acraspedote Medusæ, an order of *Hydrozoa*.

STEMMATA (στέμμα, a garland), the simple eyes of Insects, often arranged in the form of a circle on the top of the head.

STERNUM (the chest), applied to the azygos bone formed by the meeting of the visceral arches in front in most of the higher *Vertebrata*: also to the inferior pieces of the exoskeleton of *Arthropoda*.

STIGMATA (στίγμα, a mark), a synonym of *spiracula* in Insects.

STOMAPODA (more properly Stomatopoda, fr. στόμα, a mouth; πούς, foot), an order of Crustacea in which the organs of prehension retain more of the character of feet than in Decapods.

STREPSIPTERA (στρέψις, a twist; πτερόν, wing), a group of Insects with the anterior pair of wings twisted.

STROBILA, or Strobilus (στρόβιλος, a fir-cone), a chain of zooids formed by a *scolex* and the *proglottides* which have successively budded from it. The name was first applied to that of the Medusa.

SUPRA-OCCIPITAL (*supra*, above; *occiput*, i.e., *ob-caput*, the hind-head), the bone which completes the first cranial segment above, answering to the lamina of the occipital bone in Man.

SYMPHYSIS (σύμφυσις, a growing together), the union of two bones.

TÆNIA (a tape), a genus of intestinal worms, which gives its name to the order *Tæniada*, or Cestoid worms. The tape-worm, so-called, is a Strobilus formed of a scolex (the head) and proglottides (the joints).

TARDIGRADA (*tardus*, slow; *gradior*, to move), "Water-sloths," an order of *Crustacea* otherwise known as *Arctisca* or Water-bears.

TARSUS (ταρσός, the flat of the foot), the collection of small bones which form the heel or hock, and ankle; corresponding with the *carpus* in the anterior limb.

TELEOSTEI (τέλειος, perfect; ὀστέον, bone), a name given by Müller to those Fish which have completely ossified skeletons: including the Osseopterygii of Cuvier, beside the small orders Lophobranchii and Plectognathi, and corresponding with the Ctenoids (with serrated scales) and Cycloids (with smooth-edged scales) of Agassiz.

TELSON (τέλσον, a Homeric form of τέλος, end), applied to the central part of the last somite of the higher Crustacea.

TENTACULUM (*tento*, to touch), a feeler.

TEREBRATULA (dim. of *terebra*, a borer), a genus of *Brachiopoda*.

TERGUM (the back), applied to the upper segment of the exoskeleton of the somites of an Arthropod.

TEST (*testa*, a shell), applied to the chitinous covering of the *Tunicata*.

THECODONT (θήκη, a case, fr. τίθημι; ὀδούς, a tooth), having teeth implanted in distinct sockets or *alveoli*.

THORAX (θώραξ, a breastplate), applied in *Vertebrata* to the part of the trunk above the diaphragm, in Insects to the central segment formed of three consolidated somites.

THYLACINUS (θύλακος, a pouch), a genus of Marsupialia, of carnivorous habits.

THYSANURA (θύσανος, a tassel; οὐρά, a tail), a group of Insects with fringed appendages, which are attached to the end of the abdomen.

TIBIA, the shin-bone, the præaxial bone of the lower extremity answering to the *radius* in the upper.

TRABECULÆ (cranii), dim. of *trabs*, a beam, applied to the longitudinal cartilaginous bars in the embryonic skull, which enclose between them the "Sella Turcica" for the Pituitary Body.

TRACHEA (τραχεῖα, sc. ἀρτηρία, the rough windpipe), in air-breathing *Vertebrata* the tube leading to the lungs, in Insects the air-tubes which ramify throughout the body.

TREMATODA (τρῆμα, a hole; τρηματώδης is applied by Aristotle to burrowing animals), an order of Scolecida with a single opening leading to a racemose digestive system.

TRICHINA (τρίχινος, adj. of θρίξ, hair), a minute Nematoid worm parasitic in human muscle.

TRICHOCEPHALUS (θρίξ, κεφαλή) a Nematoid intestinal worm.

TRICHOPTERA (θρίξ, a hair; πτερόν, wing), an order of Insects with hairy wings.

TROCHAL disk (τροχός, a wheel), the surface around the mouth of a Rotifer, or Wheel-animalcule, on which are set the cilia whose motion produces the appearance from which it takes its name.

L

TROCHANTER (τροχαντήρ, fr. τρέχω, to turn), an outgrowth of bone from the femur which affords attachment to the muscles which rotate the thigh. In Man there are two, in the Elephant one, in *Perissodactyla* three.

TUBICOLA (*tuba*, a cylinder; *colo*, inhabit), a group of *Annelida* living in calcareous tubes which they form : e. g., *Serpulæ*.

TUNICATA (*tunica* a garment), a synonym of *Ascidioida*, Mollusks in which the shell is replaced by a chitinous test.

TURBELLARIA (*turbellæ*, pl. dim. of *turba*, a stir), a name applied (from the currents caused by their cilia?) to an order of *Scolecida*, which includes the genera *Nemertes* and *Planaria*.

TYMPANIC (τύμπανον, a drum), the bone which gives attachment to the *membrana tympani* of the ear, or its homologue.

ULNA (ὠλένη), the elbow, hence the bone of the fore-arm which forms the elbow, the *cubitus*, which answers to the *fibula* in the hind limb.

UMBILICUS (dim. of *umbo*, a boss), the navel (v. ὀμφαλός).

UNGUICULATA (*unguis*, a nail), those animals in which the dorsal part only of the digits is covered with horn, forming, if flat, a nail, if curved, a claw.

UNGULATA (*ungula*, a hoof), an order comprising all herbivorous and hoofed Mammals, including the *Pachydermata* and *Ruminantia* of Cuvier, except *Elephas* and *Hyrax*.

UNGULIGRADE (*ungula, gradior*, walk), those animals which walk on the tips of the digits only, which are always hoofed, as the horse and ruminants. Cf. DIGITIGRADE, PLANTIGRADE.

URODELA (οὐρά, tail ; δῆλος, visible), an order of *Amphibia* characterized by the tail persisting in the adult state : some are *perennibranchiate*, others *caducibranchiate*; i.e., in some the gills persist, in others they fall off after the larval stage.

VACUOLE (*vacuus*, empty), an empty space in the sarcode of *Rhizopoda* and *Infusoria*.

VAS DEFERENS, "the vessel which carries off" the seminal fluid, the duct of the testis.

VENTRICLE (*ventriculus*, dim. of *venter*, the belly), any hollow space, specially applied (as early as Cicero, who has "ventriculus cordis") to the muscular chamber of the heart which pumps blood out of that organ.

VERMES (fr. *verto*, to turn), worms, a name which with Linnæus included Insecta, Mollusca, Testacea, Zoophyta, and Infusoria. It is sometimes applied to Annuloida generally, but is better restricted to *Scolecida*.

VERTEBRA (*verto*, to turn), a joint, specially applied to those of the spinal column, and thence transferred to the small bones of which it is composed. Hence the term *Vertebrata*, since all animals belonging to this division have a more or less developed spinal column.

VESICLE (dim. of *vesica*, a bladder), applied to any sac, but specially to the *umbilical vesicle* and the *vesiculæ seminales*.

VIBRIO (*vibro*, to quiver, or vibrate), applied to minute vegetable organisms which are capable of independent movement.

VILLUS (a tuft of hair) specially applied to the vascular processes of the *chorion*, which when fully developed form the fœtal placenta.

VITELLUS (the yolk of an egg), present in all forms of ova. The *vitelline membrane* which surrounds it is otherwise called the yolk-sac.

VOMER (a ploughshare), applied to the bone which helps to form the anterior termination of the cranial axis and the *septum narium*, from its shape in Man.

GLOSSARY.

XIPHOSURA (ξίφος, sword; οὐρά, tail), a synonym of *Merostomata*, from the long sharp tail of *Limulus*.

ZEUGLODON (ζεύγλη, a yoke-strap; ὀδούς, tooth), an extinct genus of *Cetacea*, with teeth consisting of two parts united by a narrow band.

ZONARY (ζώνη, zona, a belt), that form of deciduous placenta in which the fœtal villi are arranged in a belt. A similar though broader band in a non-deciduous one is called " diffuse," or sometimes " zonular."

ZOOID (ζῷον, an animal; εἶδος, resemblance), a term applied to the individuals of compound organisms, as the polyps of a polypidom among *Cœlenterata*.

ZYGANTRUM (ζυγόν, a yoke; *antrum*, a cave), a hollow in the vertebræ of serpents by which an additional articulation is provided with the vertebra next behind.

ZYGAPOPHYSIS (ζυγόν, a yoke; ἀπόφυσις, an outgrowth), a name given to the articulating processes of vertebræ which correspond with those of Man.

ZYGOMA (ζύγωμα, fr. ζυγόν), the arch at the side of the skull formed by the jugal or yoke-bone (*os malæ* of human anatomy), articulating with the *squamosal*.

ZYGOSPHENE (ζυγόν, σφήν, wedge), a conical process on the front of the vertebræ of *Ophidia* which fits into the *zygantrum* of that next in front.

LONDON:
PRINTED BY W CLOWES AND SONS, STAMFORD STREET
AND CHARING CROSS

London, New Burlington Street,

October, 1873

A LIST

OF

MESSRS CHURCHILL'S WORKS

ON

CHEMISTRY, MATERIA MEDICA,

PHARMACY, BOTANY,

THE MICROSCOPE,

AND

OTHER BRANCHES OF SCIENCE

INDEX

	PAGE
Beasley's Pocket Formulary	ix
Do. Druggist's Receipt Book	ix
Do. Book of Prescriptions	ix
Bentley's Manual of Botany	xi
Bernays' Syllabus of Chemistry	iv
Bloxam's Chemistry	iii
Do. Laboratory Teaching	iii
Bowman's Practical Chemistry	iv
Do. Medical do.	iv
Brooke's Natural Philosophy	xv
Brown's Analytical Tables	iv
Carpenter's Microscope and its Revelations	xii
Chauveau's Comparative Anatomy	xiii
Cooley's Cyclopædia of Receipts	vi
Fayrer's Poisonous Snakes of India	xii
Fownes' Manual of Chemistry	iv
Fresenius' Chemical Analysis	v
Galloway's First Step in Chemistry	v
Do. Second do. do.	v
Do. Qualitative Analysis	v
Do. Chemical Tables	v
Greene's Tables of Zoology	xiii
Griffiths' Chemistry of the Four Seasons	v
Hardwich's Photography, by Dawson	xiv
Huxley's Anatomy of Vertebrates...	xiii
Do. Classification of Animals...	xiii
Kay-Shuttleworth's Modern Chemistry	v
Kolhrausch's Physical Measurements	xv
Lescher's Elements of Pharmacy	x
Martin's Microscopic Mounting	xii
Mayne's Medical Vocabulary	xiv
Microscopical Journal (Quarterly)...	xii
Nevins' Analysis of Pharmacopœia	ix
Ord's Comparative Anatomy	xiv
Pereira's Selecta e Præscriptis	ix
Pharmaceutical Journal and Transactions	xi
Prescriber's Pharmacopœia	xi
Price's Photographic Manipulation	xv
Proctor's Practical Pharmacy	x
Rodwell's Natural Philosophy	xv
Royle's Materia Medica...	vii
Shea's Animal Physiology	xiv
Smith's Pharmaceutical Guide	viii
Squire's Companion to the Pharmacopœia	viii
Do. Hospital Pharmacopœias	viii
Steggall's First Lines for Chemists	viii
Stowe's Toxicological Chart	x
Sutton's Volumetric Analysis	vi
Tuson's Veterinary Pharmacopœia	xi
Valentin's Inorganic Chemistry	vi
Do. Qualitative Analysis	vi
Vestiges of Creation	xiv
Wagner's Chemical Technology	vii
Wahltuch's Dictionary of Materia Medica	vii
Wilson's Zoology	xiv
Wittstein's Pharmaceutical Chemistry, by Darby	x
Year Book of Pharmacy	xi

*** *The Works advertised in this Catalogue may be obtained through any Bookseller in the United Kingdom, or direct from the Publishers, on Remittance being made.*

A LIST OF

Messrs CHURCHILL'S WORKS, &c

C. L. Bloxam

CHEMISTRY, INORGANIC and ORGANIC: With Experiments. By CHARLES L. BLOXAM, Professor of Chemistry in King's College, London; Professor of Chemistry in the Department for Artillery Studies, Woolwich. Second Edition. With 295 Engravings on Wood . . . . . . 8vo, 16s.

** It has been the author's endeavour to produce a Treatise on Chemistry sufficiently comprehensive for those studying the science as a branch of general education, and one which a student may peruse with advantage before commencing his chemical studies at one of the colleges or medical schools, where he will abandon it for the more advanced work placed in his hands by the professor. The special attention devoted to Metallurgy and some other branches of Applied Chemistry renders the work especially useful to those who are being educated for employment in manufacture.

" Professor Bloxam has given us a most excellent and useful practical treatise. His 666 pages are crowded with facts and experiments, nearly all well chosen, and many quite new, even to scientific men. . . It is astonishing how much information he often conveys in a few paragraphs. We might quote fifty instances of this." — *Chemical News.*

By the same Author

LABORATORY TEACHING: Or, Progressive Exercises in Practical Chemistry, with Analytical Tables. Second Edition. With 89 Engravings . . . . . Crown 8vo, 5s. 6d.

** This work is intended for use in the chemical laboratory by those who are commencing the study of practical chemistry. It does not presuppose any knowledge of chemistry on the part of the pupil, and does not enter into any theoretical speculations. It dispenses with the use of all costly apparatus and chemicals, and is divided into separate exercises or lessons, with examples for practice, to facilitate the instruction of large classes. The method of instruction here followed has been adopted by the author, after twenty-three years' experience as a teacher in the laboratory.

John E. Bowman and C. L. Bloxam

PRACTICAL CHEMISTRY, Including Analysis. By JOHN E. BOWMAN and C. L. BLOXAM. Sixth Edition. With 98 Engravings on Wood . . . Fcap. 8vo, 6s. 6d.

∗∗* The intention of this work is to furnish to the beginner a text-book of the practical *minutiæ* of the laboratory. The various processes employed in analysis, or which have been devised for the illustration of the principles of the science, are explained in language as simple as possible. This edition has been embellished with a large number of additional wood engravings from sketches made in the laboratory.

Also

MEDICAL CHEMISTRY. Fourth Edition, with 82 Engravings on Wood. . . . . . Fcap. 8vo, 6s. 6d.

∗∗* This work gives instructions for the examination and analysis of urine, blood, and a few other of the more important animal products, both healthy and morbid. It comprises also directions for the detection of poisons in organic mixtures and in the tissues.

— o —

Albert J. Bernays

NOTES FOR STUDENTS IN CHEMISTRY: Being a Syllabus of Chemistry and Practical Chemistry. By ALBERT J. BERNAYS, Professor of Chemistry at St. Thomas's Hospital. Fifth Edition, Revised. [Fcap. 8vo, 3s. 6d.

∗∗* A new feature is an Appendix giving the doses of the chief chemical preparations of the "Materia Medica."

"The new notation and nomenclature are now exclusively used. We notice additional notes in apparently every paragraph in the book, and a close revision of the whole."— *Scientific Opinion.*

— o —

J. Campbell Brown

ANALYTICAL TABLES for STUDENTS of PRACTICAL CHEMISTRY. By J. CAMPBELL BROWN, D.Sc. Lond., F.C.S.
[8vo, 2s. 6d.

— o —

G. Fownes

A MANUAL OF ELEMENTARY CHEMISTRY, Theoretical and Practical. BY G. FOWNES, F.R.S. Edited by Henry Watts, B.A., F.R.S. Eleventh Edition. With Wood Engravings. Crown 8vo, 15s.

Remigius Fresenius
QUALITATIVE ANALYSIS. By C. REMIGIUS FRESENIUS. Edited by Arthur Vacher. Eighth Edition, with Coloured Plate of Spectra and Wood Engravings . . . . 8vo, 12s. 6d.

By the same Author
QUANTITATIVE ANALYSIS. Edited by Arthur Vacher. Sixth Edition, with Wood Engravings . . . 8vo, 18s.

Robert Galloway
THE FIRST STEP IN CHEMISTRY: A New Method for Teaching the Elements of the Science. By ROBERT GALLOWAY, Professor of Applied Chemistry in the Royal College of Science for Ireland. Fourth Edition, with Engravings . . . Fcap. 8vo, 6s. 6d.

By the same Author
THE SECOND STEP IN CHEMISTRY: or, the Student's Guide to the Higher Branches of the Science. With Engravings.
Fcap. 8vo, 10s.

Also
A MANUAL OF QUALITATIVE ANALYSIS
Fifth Edition, with Engravings . . Post 8vo, 8s. 6d.

Also
CHEMICAL TABLES. On Five large Sheets, for School and Lecture Rooms. Second Edition . . The Set, 4s. 6d.

"We can always give praise to Mr. Galloway's educational works. They are invariably written on a system and founded on experience, and the teaching is clear, and in general complete."—*Chemical News*.

"Mr. Galloway has done much to simplify the study of chemistry by the instructive manner in which he places the principal details of the science before his readers."—*British Medical Journal*.

T. Griffiths
CHEMISTRY OF THE FOUR SEASONS : Spring, Summer, Autumn, Winter. By T. GRIFFITHS. Second Edition, with Engravings.
Fcap. 8vo, 7s. 6d.

U. J. Kay-Shuttleworth
FIRST PRINCIPLES OF MODERN CHEMISTRY. By U. J. KAY-SHUTTLEWORTH, M.P. Second Edition. Crown 8vo, 4s. 6d.

"We can recommend the book."—*Athenæum*.

"Deserving warmest commendation."—*Popular Science Rev*.

Francis Sutton

HANDBOOK OF VOLUMETRIC ANALYSIS;
or, the Quantitative Estimation of Chemical Substances by Measure applied to Liquids, Solids, and Gases. By FRANCIS SUTTON, F.C.S., Norwich. Second Edition. With Engravings . . . 8vo, 12s.

*** This work is adapted to the requirements of pure Chemical Research, Pathological Chemistry, Pharmacy, Metallurgy, Manufacturing Chemistry, Photography, etc., and for the Valuation of Substances used in Commerce, Agriculture, and the Arts.

"Mr. Sutton has rendered an essential service by the compilation of his work."—*Chemical News*.

———o———

R. V. Tuson

COOLEY'S CYCLOPÆDIA OF PRACTICAL RECEIPTS, PROCESSES, AND COLLATERAL INFORMATION IN THE ARTS, MANUFACTURES, PROFESSIONS, AND TRADES:
Including Pharmacy and Domestic Economy and Hygiène. Designed as a Comprehensive Supplement to the Pharmacopœias and General Book of Reference for the Manufacturer, Tradesman, Amateur, and Heads of Families. Fifth Edition, Revised and partly Rewritten by Professor RICHARD V. TUSON, F.C.S., assisted by several Scientific Contributors . . . . . 8vo, 28s.

"A much improved edition. . . . Long recognised as a general book of reference."—*Times*.

"The book is of considerable value for household use, as well as professional purposes, for it contains a quantity of interesting information relating to the composition of articles in common use as food and medicine."—*Pall Mall Gazette*.

"Other of the articles, as on 'brewing,' 'bread,' etc., are specimens of what cyclopædic writing should be, being concise and thoroughly exhaustive of the practical portion of the subject."—*Veterinarian*.

———o———

W. G. Valentin

INTRODUCTION TO INORGANIC CHEMISTRY.
By WM. G. VALENTIN, F.C.S., Principal Demonstrator of Practical Chemistry in the Royal School of Mines and Science Training Schools, South Kensington. With 82 Engravings . . . . 8vo, 6s. 6d.

Also

QUALITATIVE CHEMICAL ANALYSIS.
With 19 Engravings . . . . 8vo, 7s. 6d.

Also

TABLES FOR THE QUALITATIVE ANALYSIS OF SIMPLE AND COMPOUND SUBSTANCES,
both in the Dry and Wet Way. On indestructible paper . . . 8vo, 2s. 6d.

R. Wagner and W. Crookes

HANDBOOK OF CHEMICAL TECHNOLOGY. By RUDOLF WAGNER, Ph.D., Professor of Chemical Technology at the University of Wurtzburg. Translated and Edited from the Eighth German Edition, with Extensive Additions, by WILLIAM CROOKES, F.R.S.

[8vo, 25s

** The design of this work is to show the application of the science of chemistry to the various manufactures and industries. The subjects are treated of in eight divisions, as follows :—1. Chemical Metallurgy, Alloys, and Preparations made and obtained from Metals. 2. Crude Materials and Products of Chemical Industry. 3. Glass, Ceramic Ware, Gypsum, Lime, Mortar. 4. Vegetable Fibres. 5. Animal Substances. 6. Dyeing and Calico Printing. 7. Artificial Light. 8. Fuel and Heating Apparatus.

"Full and exact in its information on almost every point."—*Engineer.*
"This book will permanently take its place among our manuals."—*Nature.*

"Mr. Crookes deserves praise, not only for the excellence of his translation, but also for the original matter he has added."—*American Journal of Science and Arts.*

---o---

J. Forbes Royle and F. W. Headland

A MANUAL OF MATERIA MEDICA. By J. FORBES ROYLE, M.D., F.R.S., and F. W. HEADLAND, M.D., F.L.S. Fifth Edition, with Engravings on Wood . . . . Fcap 8vo, 12s. 6d.

** This edition has been remodelled throughout on the basis of the present edition of the British Pharmacopœia. The medicines of the British Pharmacopœia will be found arranged in natural order, the preparations described at length, and the formulæ explained. Other medicines and preparations, mentioned only in the London Pharmacopœia of 1851, are separately described and included in brackets. All remedies of value, whether officinal or not, are noticed in their place in this Manual.

"This Manual is, to our minds, unrivalled in any language for condensation, accuracy, and completeness of information."—*British Medical Journal.*

---o---

Adolphe Wahltuch

A DICTIONARY OF MATERIA MEDICA AND THERAPEUTICS. By ADOLPHE WAHLTUCH, M.D. . . . 8vo, 15s.

** The purpose of this work is to give a tabular arrangement of all drugs specified in the British Pharmacopœia of 1867. Every table is divided into six parts :—(1) *The Name and Synonyms ;* (2) *Character and Properties or Composition ;* (3) *Physiological Effects and Therapeutics ;* (4) *Form and Doses ;* (5) *Preparations ;* (6) *Prescriptions.* Other matter elucidatory of the Pharmacopœia is added to the work.

"A very handy book."—*Lancet.*

Peter Squire

COMPANION TO THE BRITISH PHARMACOPŒIA.
With Practical Hints on Prescribing; including a Tabular Arrangement of Materia Medica for Students, and a Concise Account of the Principal Spas of Europe. By PETER SQUIRE, Chemist in Ordinary to the Queen and the Prince of Wales; late President of the Pharmaceutical Society. Ninth Edition . . . . 8vo, 10s. 6d.

By the same Author

PHARMACOPŒIAS OF LONDON HOSPITALS. Second Edition . . . . Fcap 8vo, 5s.

*** Mr. SQUIRE has collected all the Formulæ used in seventeen of the principal Hospitals of London, and arranged them in groups of mixtures, gargles, &c., &c. These Formulæ were revised and approved by the medical staff of each of the Hospitals, and may therefore be taken as an excellent guide to the medical practitioner, both as to dose and best menstruum in prescribing.

---o---

J. B. Smith

PHARMACEUTICAL GUIDE. By JOHN BARKER SMITH.
Second Edition . . . . Crown 8vo.
[*In the Press.*

FIRST AND SECOND EXAMINATIONS

LATIN GRAMMAR—FRACTIONS — METRIC SYSTEM — MATERIA MEDICA — BOTANY —PHARMACY—CHEMISTRY—PRESCRIPTIONS.

---o---

John Steggall

FIRST LINES FOR CHEMISTS AND DRUGGISTS
preparing for Examination at the Pharmaceutical Society. By JOHN STEGGALL, M.D. Third Edition . . . . . 18mo, 3s. 6d.

CONTENTS

Notes on the British Pharmacopœia, the Substances arranged alphabetically.	Thermometers. Specific Gravity.
Table of Preparations, containing Opium, Antimony, Mercury, and Arsenic.	Weights and Measures. Questions on Pharmaceutical Chemistry and Materia Medica.
Classification of Plants.	

J. Birkbeck Nevins
THE PRESCRIBER'S ANALYSIS OF THE BRITISH PHARMACOPŒIA. By J. BIRKBECK NEVINS, M.D. Lond., Lecturer on Materia Medica in the Liverpool Royal Infirmary Medical School. Third Edition, Revised and Enlarged . . . . Royal 32mo, 3s. 6d.

———o———

Jonathan Pereira
SELECTA E PRÆSCRIPTIS: Containing Lists of the Terms, Phrases, Contractions, and Abbreviations used in Prescriptions, with Explanatory Notes; the Grammatical Construction of Prescriptions; Rules for the Pronunciation of Pharmaceutical Terms; a Prosodiacal Vocabulary of the Names of Drugs, &c.; and a Series of Abbreviated Prescriptions illustrating the use of the preceding terms. To which is added a Key, containing the Prescriptions in an Unabbreviated Form, with a Literal Translation for the Use of Medical and Pharmaceutical Students. By JONATHAN PEREIRA, M.D., F.R.S. Sixteenth Edition . . . . 32mo, 5s.

———o———

Henry Beasley
THE POCKET FORMULARY AND SYNOPSIS OF THE BRITISH AND FOREIGN PHARMACOPŒIAS: Comprising Standard and approved Formulæ for the Preparations and Compounds employed in Medical Practice. By HENRY BEASLEY. Ninth Edition.
[18mo, 6s.

By the same Author
THE DRUGGIST'S GENERAL RECEIPT-BOOK: Comprising a Copious Veterinary Formulary and Table of Veterinary Materia Medica; Patent and Proprietary Medicines, Druggists' Nostrums, &c.; Perfumery, Skin Cosmetics, Hair Cosmetics, and Teeth Cosmetics; Beverages, Dietetic Articles and Condiments; Trade Chemicals, Miscellaneous Preparations and Compounds used in the Arts, &c.; with useful Memoranda and Tables. Seventh Edition . . . . 18mo, 6s.

Also
THE BOOK OF PRESCRIPTIONS: Containing 3,000 Prescriptions collected from the Practice of the most eminent Physicians and Surgeons, English and Foreign. Fourth Edition . . . 18mo, 6s.

"Mr. Beasley's 'Pocket Formulary,' 'Druggist's Receipt-Book,' and 'Book of Prescriptions' form a compact library of reference admirably suited for the dispensing desk."—*Chemist and Druggist.*

F. H. Lescher
AN INTRODUCTION to the ELEMENTS of PHARMACY.
By F. HARWOOD LESCHER. Fourth Edition . 8vo, 7s. 6d.

Sec. I. MATERIA MEDICA: Characteristics of Drugs; Geographical Sources; Detection of Spurious Specimens.
II. BOTANY: Sketch of Organs, with their Functions; Groupings of the Characteristics; Natural Orders.
III. CHEMISTRY: Outline of Physics; Simple Primary Analysis; Detection of Adulterations; Poisons—Tests and Antidotes; Organic and Inorganic Chemicals.
IV. PHARMACY: Pharmacopœia; Preparations; Active Ingredients.
V. PRESCRIPTIONS: The Latin Language; Examples, with Errors and Unusual Doses; Tables of Doses.
VI. PRACTICAL DISPENSING: Groupings of Strengths of Solutions; Emulsions; Pills, &c.; Changes in Mixtures.

—o—

B. S. Proctor
LECTURES ON PRACTICAL PHARMACY.
By BARNARD S. PROCTOR, Lecturer on Pharmacy at the College of Medicine, Newcastle-on-Tyne. With 43 Wood Engravings . 8vo, 12s.

₊ The object of the writer is to assist earnest Students by indicating the direction and manner in which the study of Pharmaceutical subjects should be pursued; attention being principally directed to such points as are not included in the usual Manuals of Chemistry and Materia Medica. The object is divided into—

Abstract Processes: Drying, Grinding, Solution, Diffusion, Filtration, etc.
Official Processes.
Extempore Processes: Dispensing Mixtures, Pills, Plasters, Ointments, etc. Reading difficult Autographs, illustrated with lithographic fac-similes.
Official Testing. Notes on the Qualitative and Quantitative Systems of the Pharmacopœia.
Pharmacy of Special Drugs, being Studies of Cinchona, Opium, Aloes, and Iron.

—o—

William Stowe
A TOXICOLOGICAL CHART, Exhibiting at one view the
Symptoms, Treatment, and Mode of Detecting the Various Poisons, Mineral, Vegetable, and Animal. To which are added concise Directions for the Treatment of Suspended Animation. By WILLIAM STOWE, M.R.C.S.E. Thirteenth Edition . . . . Sheet, 2s.; Roller, 5s.

—o—

G. C. Wittstein
PRACTICAL PHARMACEUTICAL CHEMISTRY: An
Explanation of Chemical and Pharmaceutical Processes; with the Methods of Testing the Purity of the Preparations, deduced from Original Experiments. By Dr. G. C. WITTSTEIN. Translated from the Second German Edition by STEPHEN DARBY . . . . . . 18mo, 6s.

"It would be impossible too strongly to recommend this work to the beginner, for the completeness of its explanations, by following which he will become well grounded in practical chemistry."—*From the Introduction by Dr. Buchner.*

THE PRESCRIBER'S PHARMACOPŒIA.: The Medicines arranged in Classes according to their Action, with their Composition and Doses. By A PRACTISING PHYSICIAN. Fifth Edition.
[Fcap 16mo, cloth, 2s. 6d.; roan tuck, 3s. 6d·

———o———

THE PHARMACEUTICAL JOURNAL AND TRANSAC-
TIONS. Published weekly . . . . Price 4d.

———o———

THE YEAR-BOOK OF PHARMACY: Containing the Proceedings at the Yearly Meeting of the British Pharmaceutical Conference, and a Report on the Progress of Pharmacy, which includes notices of all Pharmaceutical Papers, new Processes, Preparations, and Formulæ published throughout the world. Published annually . . 8vo, 7s. 6d.

———o———

R. V. Tuson

A PHARMACOPŒIA, INCLUDING THE OUTLINES OF MATERIA MEDICA AND THERAPEUTICS, for the Use of Practitioners and Students of Veterinary Medicine. By RICHARD V. TUSON, F.C.S., Professor of Chemistry and Materia Medica at the Royal Veterinary College. Second Edition . . . . *In the press.*

"Not only practitioners and students of veterinary medicine, but chemists and druggists will find that this book supplies a want in veterinary literature."—*Chemist and Druggist*.

———o———

Robert Bentley

A MANUAL OF BOTANY: Including the Structure, Functions, Classifications, Properties, and uses of Plants. By ROBERT BENTLEY, F.L.S., Professor of Botany, King's College, and to the Pharmaceutical Society. Third Edition, with 1,127 Wood Engravings.
[Crown 8vo. *Just ready.*

"As the standard manual of botany its position is undisputed."—*Chemist and Druggist*.

W. B. Carpenter

THE MICROSCOPE AND ITS REVELATIONS. By W. B. CARPENTER, M.D., F.R.S. Fifth Edition, with more than 500 Wood Engravings . . . . Crown 8vo. *Just ready.*

** The author has aimed to combine within a moderate compass that information in regard to the use of his instrument and its appliances, which is most essential to the working microscopist, with such an account of the objects best fitted for his study as may qualify him to comprehend what he observes, and thus prepare him to benefit science, whilst expanding and refreshing his own mind.

―――o―――

J. H. Martin

A MANUAL OF MICROSCOPIC MOUNTING; with Notes on the Collection and Examination of Objects. By JOHN H. MARTIN, author of "Microscopic Objects." With upwards of 100 Engravings.

[8vo, 7s. 6d.

** The aim of this work is to supply the student with a concise manual of the principles of microscopic mounting, and to assist his progress in the manual dexterity, as far as illustrations and words render it possible, necessary in their application.

―――o―――

THE QUARTERLY JOURNAL OF MICROSCOPICAL SCIENCE. (Established in 1852.) Edited by Dr. J. F. PAYNE, Demonstrator of Morbid Anatomy, and Assist.-Physician at St. Thomas's Hospital; Mr. E. RAY LANKESTER, Fellow of Exeter College, Oxford; and W. T. THISELTON DYER, Professor of Botany to the Royal Horticultural Society.

[Annual Subscription, 16s.; Single Numbers, 4s.

** The Memoirs are, when needful, illustrated by Lithographic Plates, many of which are Coloured. The Journal contains, in addition, Notes and Memoranda, Reviews of Books, Quarterly Chronicle, and Proceedings of Societies.

―――o―――

J. Fayrer

THE THANATOPHIDIA OF INDIA; being a Description of the Venomous Snakes of the Indian Peninsula. With an Account of the Influence of their Poison on Life, and a Series of Experiments. By J. FAYRER, M.D., C.S.I., Honorary Physician to the Queen; late President of the Asiatic Society of Bengal. Second Edition, with 31 Plates (28 Coloured) . . . . Folio, 7*l.* 7s.

A. Chauveau and G. Fleming

CHAUVEAU'S COMPARATIVE ANATOMY OF THE DOMESTICATED ANIMALS.
Translated from the Second French Edition, and Edited by GEORGE FLEMING, F.R.G.S., Veterinary Surgeon, Royal Engineers; Author of "Travels on Horseback in Mantchu Tartary," "Horse-shoes and Horse-shoeing," "Animal Plagues," etc. With 450 Engravings on Wood . . . . 8vo, £1 11s. 6d.

"The want of a text-book on the Comparative Anatomy of the Domesticated Animals has long been felt. The present work is the fruit of a desire to fill a void in medical literature which has always existed, so far as the English language is concerned. The care and attention with which hippotomy has been cultivated on the Continent are illustrated by every page in M. Chauveau's work. . . . If we compare the description, say of the arteries of the head and neck of the horse, as given in Chauveau's work, with the elaborate description given in Quain or Ellis of the same arteries in man, we shall find that in minuteness of detail the anthropotomist has been very closely rivalled by the hippotomist. . . In taking leave of this book we may congratulate Mr. Fleming on the completion of so great and useful an undertaking. He has translated his author into excellent scientific English, and his contributions (which in the text are placed between brackets) are proof of the large amount of study and research he has given to make the book as complete as possible. He has not only produced a most valuable—and, in fact, the only—anatomical text-book for the veterinary student, but he has given us a work to be prized by every scientific man who wishes to become acquainted with the anatomy of the higher vertebrata."—*Medical Times and Gazette*, May 10, 1873.

"This is a valuable work, well conceived and well executed by the authors, MM. Chauveau and Arloing, and well translated by Mr. Fleming. Altogether the work reminds us very much of Quain and Sharpey's, where the histological part in the latter intercalated with the systematic; and this is giving it no slight praise. We have compared M. Chauveau's description of the bones and other organs, where practicable, with those of Owen, Huxley, Flower, and other English writers, and find that they are in general very accurate and good. . . . The illustrations are very numerous, and Mr. Fleming has introduced a large number that are not contained in the original work."—*Lancet*, May 31, 1873.

—o—

J. Reay Greene

TABLES OF ZOOLOGY:
indicating the Tribes, Sub-Orders, Orders, and Higher Groups of the Animal Kingdom, for Students, Lecturers, and others. By J. REAY GREENE, M.D., Professor of Natural History in the Queen's University in Ireland. Three large sheets, 7s. 6d. the set; or, mounted on canvas, with roller and varnished . . 18s.

*** These Tables have been carefully prepared in accordance with the present state of science, and with a view to remove the difficulties which arise from the various opinions held by different zoologists.

—o—

T. H. Huxley

A MANUAL OF THE ANATOMY OF VERTEBRATED ANIMALS.
By Prof. HUXLEY, LL.D., F.R.S. With numerous Engravings. . . . . [Fcap. 8vo, 12s.

By the same Author

INTRODUCTION to the CLASSIFICATION of ANIMALS.
With Engravings . . . . 8vo, 6s.

W. M. Ord
NOTES ON COMPARATIVE ANATOMY: a Syllabus of
a Course of Lectures delivered at St. Thomas's Hospital. By WILLIAM MILLER ORD, M.B. Lond., M.R.C.P., Assistant-Physician to the Hospital, and Lecturer in its Medical School . . . . . Crown 8vo, 5s.

"Compact, lucid, and well arranged. These Notes will, if well used, be valuable to learners, perhaps still more so to teachers."—*Nature.*

"We have gone through it carefully, and we are thoroughly satisfied with the manner in which the author has discharged his task."—*Pop. Science Review.*

———o———

John Shea
A MANUAL OF ANIMAL PHYSIOLOGY. With Appendix
of Examination Questions. By JOHN SHEA, M.D., B.A. Lond. With numerous Engravings . . . . . . Fcap. 8vo, 5s. 6d.

———o———

VESTIGES of the NATURAL HISTORY OF CREATION.
With 100 Engravings on Wood. Eleventh Edition . Post 8vo, 7s. 6d.

———o———

Andrew Wilson
THE STUDENT'S GUIDE TO ZOOLOGY.
By ANDREW WILSON, Author of "Elements of Zoology," and Lecturer on Zoology, Edinburgh. With Engravings . . . *In November*

———o———

R. G. Mayne
MEDICAL VOCABULARY: an Explanation of all Names,
Synonymes, Terms, and Phrases, used in Medicine and the Relative Branches of Medical Science, giving their correct Derivation, Meaning, Application, and Pronunciation. Intended specially as a Book of Reference for the Young Student. Third Edition . . . Fcap 8vo, 8s. 6d.

"We have referred to this work hundreds of times, and have always obtained the information we required . . . Chemical,

Botanical, and Pharmaceutical Terms are to be found on almost every page."—*Chemist and Druggist.*

———o———

G. Dawson
A MANUAL OF PHOTOGRAPHY. By GEORGE
DAWSON, M.A., Ph.D., Lecturer on Photography in King's College, London. Eighth Edition, with Engravings . . Fcap 8vo, 5s. 6d.

"The new edition of this excellent manual, which is founded on and incorporates as much of Hardwich's 'Photographic Chemistry' as is valuable in the present further advanced stage of the art, retains its position as the best work on the subject for amateurs, as well as professionals. The

many new methods and materials which are so frequently being introduced, make it essential that any book professing to keep up to the times must be frequently revised, and Dr. Dawson has in this work presented the subject in its most advanced position."—*Nature*, May 29, 1873.

Lake Price
A MANUAL OF PHOTOGRAPHIC MANIPULATION.
By LAKE PRICE. Second Edition, Revised and Enlarged, with numerous Engravings . . . . Crown 8vo, 6s. 6d.

*** Amongst the Contents are the Practical Treatment of Portraits—Groups in the Studio—Landscapes—Groups in Open Air—Instantaneous Pictures—Animals—Architecture—Marine Subjects—Still Life—Copying of Pictures, Prints, Drawings, Manuscripts, Interiors—Stereoscopy in Microphotography, &c., and Notices of the last Inventions and Improvements in Lenses, Apparatus, &c.

"In these days, when nearly every intelligent person can, after a few weeks, master the manipulatory details of our art-science, attention to the artistic treatment of subjects is a matter for the serious considera-' tion of the Photographer; and to those who desire to enter on this path, Mr. LAKE PRICE. in the volume before us, proves himself to be ' a guide, philosopher, and friend.' "—*The British Journal of Photography.*

——o——
C. Brooke
THE ELEMENTS OF NATURAL PHILOSOPHY. By
CHARLES BROOKE, M.B., M.A., F.R.S. Based on the Work of the late Dr. GOLDING BIRD. Sixth Edition, with 700 Engravings on Wood.
[Fcap 8vo, 12s. 6d.

CONTENTS
1, Elementary Laws and Properties of Matter: Internal or Molecular Forces— 2, Properties of Masses of Matter: External Forces—3, Statics—4, The Mechanical Powers, or Simple Machines—5, Principles of Mechanism—6, Dynamics—7, Hydrostatics—8, Hydrodynamics—9, Pneumatics—10, Acoustics—11, Magnetism; Diamagnetism—12, Franklinic Electricity—13, Voltaic Electricity—14, Electro-Dynamics— 15, Electro-Telegraphy—16, Thermo-Electricity—17, Organic Electricity—18, Catoptrics and Dioptrics—19, Chromatics—20, Optical Instruments—21, Polarised Light— 22, Chemical Action of Light: Photography—23, Thermics—24, Radiant Heat.

——o——
G. F. Rodwell
NOTES ON NATURAL PHILOSOPHY.
By G. F. RODWELL, F.R.A.S., Lecturer on Natural Philosophy in Guy's Hospital, Science Master in Marlborough College. With 48 Wood Engravings . . . . . Crown 8vo, 5s.

"As an introductory text-book for this Examination [the Preliminary Scientific (M.B.) of the University of London], it is quite the best one we have seen . . The Notes' chiefly consist of lucid and concise definitions. and everywhere bristle with the derivations of scientific terms."—*Nature.*
"A well-arranged and carefully-written condensation of the leading facts and principles of the chief elements of Natural Philosophy."—*Chemical News.*

——o——
F. Kohlrausch
AN INTRODUCTION TO PHYSICAL MEASUREMENTS,
With Appendices on Absolute Electrical Measurement, etc. By Dr. F. KOHLRAUSCH. Translated from the Second German Edition by T. H. WALLER, B.A., B. Sc., and H. R. PROCTER, F.C S. With Engravings.
[8vo, 12s.

THE following CATALOGUES issued by Messrs CHURCHILL will be forwarded post free on application:

1. *Messrs Churchill's General List of 400 works on Medicine, Surgery, Midwifery, Materia Medica, Hygiene, Anatomy, Physiology, Chemistry, &c., &c.*

2. *Selection from Messrs Churchill's General List, comprising all recent works published by them on the Art and Science of Medicine.*

3. *A descriptive List of Messrs Churchill's works on Chemistry, Pharmacy, Botany, Photography, and other branches of Science.*

4. *Messrs Churchill's Red-Letter List, giving the Titles of forthcoming New Works and New Editions.*
[Published every October.]

5. *The Medical Intelligencer, an Annual List of New Works and New Editions published by Messrs J. & A. Churchill, together with Particulars of the Periodicals issued from their House.*

[Sent in January of each year to every Medical Practitioner in the United Kingdom whose name and address can be ascertained. A large number are also sent to the United States of America, Continental Europe, India, and the Colonies.]

MESSRS CHURCHILL have concluded a special arrangement with MESSRS LINDSAY & BLAKISTON, OF PHILADELPHIA, in accordance with which that Firm will act as their Agents for the United States of America, either keeping in Stock most of Messrs CHURCHILL'S Books, or reprinting them on Terms advantageous to Authors. Many of the Works in this Catalogue may therefore be easily obtained in America.

www.ingramcontent.com/pod-product-compliance
Lightning Source LLC
Chambersburg PA
CBHW020305170426
43202CB00008B/503